Resisting Global Toxics

Urban and Industrial Environments

Series editor: Robert Gottlieb, Henry R. Luce Professor of Urban and Environmental Policy, Occidental College

For a complete list of books published in this series, please see the back of the book.

Resisting Global Toxics

Transnational Movements for Environmental Justice

David Naguib Pellow

The MIT Press
Cambridge, Massachusetts
London, England

MIT Press books may be purchased at special quantity discounts for business or sales promotional use. For information, e-mail special_sales@mitpress.mit.edu or write to Special Sales Department, The MIT Press, 55 Hayward Street, Cambridge, MA 02142.

This book was set in Sabon by Graphic Composition, Inc.
Printed on recycled paper and bound in the United States of America.

Library of Congress Cataloging-in-Publication Data

Pellow, David Naguib, 1969–
Resisting global toxics : transnational movements for environmental justice / David Naguib Pellow.
p. cm. — (Urban and industrial environments)
Includes bibliographical references and index.
ISBN 978-0-262-16244-9 (hardcover : alk. paper) — ISBN 978-0-262-66201-7 (pbk. : alk. paper)
1. Environmental justice. 2. Hazardous wastes—Developing countries. 3. Environmental protection—International cooperation. I. Title.
GE220.P45 2007
363.72'87091724—dc22
2006035516

10 9 8 7 6 5 4

Contents

Acknowledgments

This project took nine years from start to finish, and there are many people who supported my efforts during that time. I am grateful for outstanding research assistance from: Katrina Behrend, Miya Saika Chen, Sabrina Cohen, Maya Dehner, Sabrina Eileen Hodges, Sabrina Khandwalla, Denise Khor, Anthony Kim, Sanna King, Rebecca Kinney, Emily Kohl, Jason Kai Lum, Mia Monreal, Errol Schweizer, Theresa Cenidoza Suarez, Angela Trujillo, Alex Urquhart, and Traci Brynne Voyles. I want to thank my colleagues and friends who labor full time in the trenches doing environmental justice, labor rights, and human rights work and who have been so generous with their time and energy. They include: the boards and staffs of the Center for Urban Transformation, the Environmental Health Coalition, the European Roma Rights Center, the Global Anti-Incinerator Alliance or the Global Alliance for Incinerator Alternatives, Global Greengrants Fund, Global Response, Greenpeace (USA and International), groundWork (USA and South Africa), the International Campaign for Responsible Technology, the International Rivers Network, Livaningo, One Earth One Justice, People for Community Recovery, Pesticide Action Network Asia and the Pacific, the Santa Clara Center for Occupational Safety and Health, the Silicon Valley Toxics Coalition, Taiwan Environmental Action Network, and WorkSafe! Specific persons in these organizations and other groups who have been particularly generous and patient with me include: Ravi Agarwal, Chris Allan, Kenny Bruno, Arden Buck, Leslie Byster, Shenglin Chang, Mariella Colven, Francis dela Cruz, Jorge Emmanuel, Connie Garcia, Bob Golten, Monique Harden, Amanda Hawes, Von Hernandez, Nity Jayaraman, Cheryl and Hazel Johnson, Heeten Kalan,

Erin King, Anabela Lemos, Jenny Lin, Patrick McCully, Bob Moran, Jorge Osuna, Bobby Peek, Jim Puckett, Romeo Quijano, Rebecca McLain, Marni Rosen, Raquel Sancho, Mageswari Sangaralingam, Sathyu Sarangi, David Silver, Amelia Simpson, Paul Strasburg, Diane Takvorian, Chet Tchozewski, Mily Trevino, Wenling Tu, Kishore Wankhade, Emily Young, and Roy Young.

I am fortunate to have a great network of colleagues and friends at various universities, whom I must acknowledge for their guidance during these past several years: Ann Briggs Addo, Julian Agyeman, Theresa Aitchison, Roberto Alvarez, Alexios Antypas, Mark Appelbaum, Bernard Beck, Larry Boehm, Charles Briggs, Robert Bullard, Beth Schaefer Caniglia, Marsha Chandler, Angana Chatterji, Dan Cress, Shalanda Dexter, Liam Downey, Paul Drake, Riley Dunlap, Melanie DuPuis, Yolanda Escamilla, Yen Le Espiritu, Gary Alan Fine, Dorian Fougeres, Ross Frank, R. Scott Frey, Patrick Gillham, Jackie Griffin, Nikolas Heynen, Greg Hooks, Evelyn Hu-DeHart, Lynn Hudson, Lori Hunter, James Jacobs, Eric Klinenberg, Brian Klocke, Paul Lubeck, Marta Maldonado, Richard Marciano, Jorge Mariscal, John Marquez, José Martinez, Glenna Matthews, Jane Menken, Carolyn Merchant, Dennis Mileti, David Miller, Natalia Molina, Harvey Molotch, Karen Moreira, Juanita Pasallo, Nancy Peluso, Keith Pezzoli, Jane Rhodes, Bill Runk, Lisa Sanchez, Hiram Sarabia, Lindy Shultz, Richard Shapiro, Denise Ferreira da Silva, Chad Smith, Tamara Steger, Andrew Szasz, Julie Sze, Celeste Watkins, Peter Wissoker, Diana Wu, Shinjoung Yeo, Joseph Yue, and Ana Celia Zentella.

Special thanks to the fellows at the Robert Wood Johnson Foundation Scholars in Health Policy Research Program at the University of California at Berkeley's School of Public Health, the participants in the Urban Studies Workshop at New York University's Department of Sociology, the participants at Washington State University's Department of Sociology's Workshop on Environmental Sociology, the participants at Northwestern University's Department of Sociology's Colloquium Series, the participants at the Workshop on Environmental Politics at the University of California, Berkeley, and the participants at the University of California, Santa Cruz's Department of Sociology's Colloquium Series. So many of you gave me helpful feedback and criticism on early drafts of papers that evolved into this book.

The following institutions and programs graciously provided funding for various stages of this research: the Robert Wood Johnson Foundation's Scholars in Health Policy Research Program; the Trust for Mutual Understanding; University of California Institute for Labor and Employment; University of California, San Diego's Faculty Mentor Program and McNair Scholars Program; the Council on Research and Creative Work, the Dean's Fund for Excellence, the Implementation of Multicultural Perspectives and Approaches in Research and Teaching Award Program, and the Undergraduate Research Opportunity Program, all at the University of Colorado.

A very special thanks to my close academic collaborators Robert J. Brulle, Kenneth Gould, Lisa Sun-Hee Park, Allan Schnaiberg, David Sonnenfeld, and Adam Weinberg—I would be nowhere without you. I also owe a huge debt of gratitude to my good friends and activist collaborators Annie Leonard, Paula Palmer, Ted Smith, and Orrin Williams—my heroes! And many thanks to my editors, Clay Morgan and Robert Gottlieb, who have given me such extraordinary support for many years. And finally, my thanks to the great artist and activist, Joaquin "Quino" McWhinney, for demanding that university scholars "write books that people can actually read!"

1

Environment, Modernity, Inequality

One morning in 1987, on the Southeast Side of Chicago, several African American environmental justice activists, along with their white allies from a large environmental organization, engaged in an act of civil disobedience against an incinerator operator located in the community. They coordinated a lockdown, chaining themselves to vehicles placed in the path of trucks transporting hazardous materials for incineration. The activists held their ground for several hours, in defiance of the company—one of dozens of highly polluting operations in this African American community. By the end of the day, the coalition had turned away no fewer than fifty-seven waste trucks.[1] Hazel Johnson, founder of the environmental justice group People for Community Recovery (PCR), recounted this story on several occasions and was always proud of the fact that her group had led the demonstration. Indeed, this was a remarkable mobilization and impressive act of resistance from within a small, low-income community of color.

This case was important not only because it reflected the power of local community-based activism against environmental inequality—the heavy burden of toxic pollution imposed on poor communities and people of color—but also because it involved a multiracial collaboration between grassroots social movement organizations (SMOs) and because all actors involved in this seemingly local struggle have strong transnational ties. Following this action, PCR staff members would soon travel to and work with activists from Brazil, Nigeria, Puerto Rico, South Africa, and various Native American communities. Their collaborating environmental organization in the lockdown action was Greenpeace, a global nongovernmental organization (NGO) with offices, personnel, and campaigns in dozens of nations, and the company being targeted in this action was ChemWaste, a hazardous materials subsidiary of Waste Management (WMX). At the time, WMX was the world's largest waste management firm, with revenue in the billions of dollars and operations on several continents. In this light, what appeared to be a conflict between activists and a company in one small community was also reflective of how many environmental justice struggles are simultaneously local and global and how this case foreshadowed the growing globalization of the environmental justice movement.

PCR was born out of a conflict over health and environmental justice that had deep local roots and an international reach. The organization faced insensitivity from local elected officials and government agencies whose charge was to protect the environment and public health. Local activists succeeded by building a support base at home and outside their community to raise the stakes for the offenders, who now faced formidable opposition. PCR's battle against ChemWaste that day occurred at the same moment many global South communities were being targeted by waste firms and chemical producers. Indeed, in some ways, the successes of groups like PCR often placed greater pressure on communities in the global South, where people have command of fewer resources.[2] Transnational environmental justice offenses require transnational responses.

Four Interventions

This book is an exploration of the export of hazardous waste (through trading and dumping) to poor communities and communities of color around the world and charts the mobilization of transnational environmental justice movement networks to document and resist these practices. Building on the work of scholars of environmental justice studies, environmental sociology, social movement theory, and race theory, I argue that the practice of waste dumping across national borders is a form of transnational environmental inequality and is reflective of unequal, and deeply racialized, relations between and within global North and South communities that transnational social movement networks are combating with great ingenuity. I use the term *global South* mainly as a social—rather than strictly geographic—designation meant to encompass politically and economically vulnerable communities. Thus, while I sometimes use the terms *global North nations* and *global South nations,* I also include communities of color and poor communities in industrialized nations within the "South" designation (or what some observers call "the South of the North"[3]) and privileged communities in poor nations within the "North" designation (or the "North of the South"[4]). In this way, we complicate the basic spatial, geographic, and cultural dichotomies implied by the global North/global South binary. Thus, we can also draw clear connections between environmental inequalities facing communities domestically and internationally, because these processes are tightly linked, have common roots, and spawn similar responses from citizens and activists.[5]

Every year, northern nations and corporations produce millions of tons of toxic waste from industry, consumers, municipalities, state institutions, computers and electronics products, and agricultural practices. These hazards directly and indirectly contribute to high rates of human (and nonhuman) morbidity and mortality and to ecosystem damage on every continent and ocean system. As long as societies produce this waste, it must go somewhere, but few communities welcome these poisons within their borders. How then do so many communities across the globe end up playing host to these deadly substances? In what ways are they fighting back? In what sense do these conflicts reflect larger historical, economic, and social realities between the global North and South and between groups within these societies? How then do we theorize race, class, nation, and the environment in a transnational context? These are some of the questions I explore in this study, through four *interventions,* or critical contributions to a number of scholarly literatures.

Intervention One: Transnational Waste Trading and Dumping

The first intervention concerns the literatures on the transnational waste trade. These studies mainly focus on the trend of waste shifting from North to South, how these practices reflect global economic inequalities among nations,[6] and how NGOs have participated in shaping multilateral agreements regulating or banning these practices. Drawing on international relations, legal studies, and world systems theory, the first series of scholarly studies of the international waste trade laid the foundation for new questions. For example, this literature paid little attention to how one might conceptually frame this problem in ways that speak to ongoing debates in race and ethnic studies, environmental sociology, or social movement theory. That is, referring to transnational hazardous waste dumping as environmental racism without linking it to theories of racism—as scholars have generally done—is limiting. Similarly, examining transnational waste dumping without reference to theories of environment and modernity leaves out critical questions of the role of capital and state formation that facilitate this process. Finally, if we focus mainly on the work that environmental movements do within formal multilateral policymaking bodies, we overlook the more routine, informal, grassroots efforts of movements within the very communities targeted for toxic waste dumping and how they build strategic ties to NGOs outside their nations, thus creating influential movement networks. Most important, a focus on movement networks linked to the target sites of waste dumping challenges the perception in the literature that citizens in these nations are powerless victims.

Intervention Two: Environmental Justice Studies and Late Modernity

The second intervention is my effort to place environmental justice studies in a broader framework that considers the toxic nature of late modernity itself. That is, I seek to build on environmental justice studies' focus on how environmentally unequal practices harm vulnerable populations to explore more fully how these inequalities reveal something deeply problematic about the relationship between modernity and the environment.[7] I attend to this concern primarily through a critical reading of theories of environment-society dynamics, focused on the impacts of industrialization on ecosystems and social systems. Ecological modernization theory contends that industries are integrating sustainability goals into their core operations, leading to measurable improvements in ecosystems, while Ulrich Beck's risk society and Allan Schnaiberg's treadmill of production theories essentially argue the opposite—that is, market economies and governments are increasingly socially and ecologically unsustainable in this late modern era.[8] I focus on the extent to which these frameworks hold explanatory power. I find that ecological modernization theory works when coupled with global environmental inequalities, because the latter facilitate the shift of negative environmental externalities from privileged northern communities to poorer, southern communities with people of color majorities.[9] But ultimately the risk society and treadmill of production perspectives are more productive for understanding the global dynamics of political economy and toxicity in modern nation-states. I address these questions in greater depth later in this chapter.

Intervention Three: Environment, Race, Class, and Nation

The third intervention is the need to unpack the relationship among modernity, the environment, race, class, and nation. Building on the second intervention, I begin with the claim that the basic functions of industrialized societies (primarily in the global North) involve the production of both intense ecological harm and extensive social hierarchies (primarily by race, class, gender, and nation).[10] The intersection of social inequalities with ecological harm produces environmental inequality both domestically (within nations) and on a transnational scale (between northern and southern nations and regions).[11] Ecological disorganization and environmental inequality and racism are therefore fundamental to the project of modern nation building. The extension of unearned privileges to certain groups and unjust disadvantages to "others" in the context of the systemic manipulation and exploitation of nature is a defining feature of modern nation-states. This dynamic alters our understanding of nation and of the nature of racism and class domination. Not only is the state's existence predicated on the manipulation of the natural environment and the devaluation of people of color, indigenous peoples, and the poor, but the practices of racism and class domination themselves must be redefined as the domination of people and their environment. Thus, in linking theories of modernity, race, class, and the environment, I contend that the exploitation of humans and the environment is a unified practice and is the foundation of racism and class inequalities, a cornerstone of modern nation building itself. These observations build on the work of environmental historians,[12] environmental sociologists,[13] and race theorists[14] who wrestle with the relationships between environmental harm and modernity and the relationships between racism and modernity but do so separately. I intend to join these parallel conversations. More specifically, I ask how we can productively merge environmental justice studies and race theories.

One question in that regard concerns the current discourse and debates over whether today's racial common sense adheres more closely to a colorblind (or postracial) model versus a more visible, structured racial reality.[15]

And while racism may operate differently across national boundaries, the rhetoric of color-blind race relations has become a global common sense, as we see it deployed in the United States, Mexico, Puerto Rico, France, and Australia, to name only a few places.[16] Racism for many theorists has become centered around the question of colorblindness largely because in this late modern era, many of the technologies of racism enjoy greater invisibility. Toward that end, I also grapple with the ideas of racial justice activists and critical race theorists who have consistently described racism through the use of the metaphor of poison or a toxic practice. I develop this metaphor to reveal how theories of racism and theories of environmental justice can be integrated to demonstrate both the symbolic and material toxicity of racism and class domination. Through the concept of toxicity, we understand how racism and class inequalities can simultaneously operate invisibly and quite blatantly, the way risks move through Ulrich Beck's risk society and the way power moves through Allan Schnaiberg's treadmill of production. I argue that racism and class inequalities reinforce each other and become more visible when vulnerable communities confront environmental harm. I address these issues in greater depth later in this chapter.

Intervention Four: Social Movements, Nation-States, and Capital

The fourth intervention I pursue builds on the first three. Here I engage in a conversation with scholars regarding the nature and efficacy of social movements, particularly transnational social movement organizations (TSMOs) and transnational movement networks, and their efforts to combat the environmental inequalities associated with the project of nation building and the globalization of market economies. In so doing, these activist organizations and their networks necessarily work at multiple geographic, geopolitical, symbolic, and cultural scales. This includes confronting state authorities at the local, regional, national, and international levels, as well as efforts to shape and enforce international conventions, treaties, and multilateral environmental agreements; it also involves policy work and negotiations with international financial institutions (IFIs) such as the World Bank and the International Monetary Fund (IMF) and confronting transnational corporations (TNCs) for their social and environmental practices. Local and transnational SMOs featured in this study strategically employ their energies and target authorities at all scales, depending on which point of access is likely to yield the greatest political payoff. This "venue shopping" is particularly useful when infrastructure or political support is insufficient at any one level.[17] For example, one case features NGOs in the Philippines that invoke the Philippine constitution, the U.S. Environmental Protection Agency's (USEPA) regulatory framework, and the Stockholm Convention on the Elimination of Persistent Organic Pollutants to frame and bolster their argument for the introduction of new waste management legislation in their nation. This tactic was successful and indicates that transnational movements for environmental justice have become quite sophisticated at combating global environmental inequalities in numerous political spaces, using multiple tactics. This is also indicative of a grounded form of global citizenship that is illuminated in practice, through the engagement of a range of institutions. In my effort to draw on the work of social movement scholars, I argue that these researchers might pay more attention to the ways in which national and transnational political opportunity structures are intensely racialized, classed, and routinely shaped by TNCs. Theoretically, this perspective seeks an integration of social movement theory with theories of racism and environmental justice studies around several questions: What are the targets of social movements in a global political economy? What tactics and strategies are movements developing in a globalizing world, and how effective are they? How are transnational political and economic processes racialized and classed, and what is the significance for social movements fighting for justice in poor communities and communities of color? These questions imply critiques of social movement theory, as leading scholars in that field have yet to raise these concerns.[18] I address these questions in greater depth in the next chapter.

Together, these four interventions converge to make contributions to the study of environmental sociology; theories of race, class, and modernity; environmental justice studies; and social movement theory.

The next sections introduce the primary subject of concern in this study: the global trade and shifting of hazardous wastes from communities in the global North to those in the global South. I then interrogate this problem through the lenses of environmental justice studies, the risk society, the treadmill of production, and ecological modernization.

The Global Waste Trade

Since the end of World War II, industrialized nations have generated increasing volumes of hazardous chemical wastes, a result of technological developments across all industry sectors and a culture of increasing acceptance of risk in late modernity. Today it is estimated that nearly 3 million tons of hazardous waste from the United States and other industrialized nations cross international borders each year. Of the total volume of hazardous waste produced worldwide, 90 percent of it originates in industrialized nations. Much of this waste is being shipped from Europe, the United States, and Japan to nations in Latin America, the Caribbean, South and Southeast Asia, and Africa. This is a global problem paralleling the domestic struggle against environmental inequality within the United States. And as with all other forms of racism and inequality, it is historically contingent on forces that are in constant tension and therefore change over time. In fact, the problems of the global environmental racism and inequality have intensified over the past two decades, revealing how fluid and dynamic social hierarchies can be.[19]

There are four principal reasons for this shift of toxic burdens to the global South. First is the exponential increase in the production of hazardous waste and the emergence of more stringent environmental regulations in industrialized nations. These changes have increased the costs of waste treatment and disposal in the North, which are magnitudes greater than in most southern nations. Similarly, the typical legal apparatus found in industrialized nations is much more burdensome when compared to the lax regulatory regimes in many nations in the South, which allow dumping at a fraction of the cost. This is due partly to a comparatively more influential environmental movement sector in industrialized nations, which has successfully produced a regulatory structure that provides a minimal level of oversight over polluting firms. The unintended consequence of this environmentalist "success" in the North is to provide an incentive for the worst polluters to seek disposal sites beyond national borders.[20]

A second factor pushing hazardous waste beyond northern borders is the widespread need for fiscal relief among southern nations. This need—rooted in a long history of colonialism and contemporary loan and debt arrangements between southern and northern nations—often leads government officials in the South to accept financial compensation in exchange for permission to dump chemical wastes in their borders.[21] Many observers (for example, economists and business leaders in northern countries) have described these transactions as "economically efficient," while others (for example, African elected officials and environmentalists in the South) prefer the term *garbage imperialism.*[22]

The third driving force behind the international export of hazardous materials is the seemingly inexorable power of economic globalization, which has a logic that dictates that industries must cut costs and increase profits or simply fail.[23] Economic globalization allows and requires firms to access global (consumer and commercial) markets and labor forces, increase automation, and improve efficiencies in a twenty-four-hour economy that is more interdependent than ever before. The same logic applies to industries that manage the hazardous waste that market economies produce: they must access markets and buyers where the prices result in increasing their profits and reducing their costs. This means those wastes will be traded and dumped in nations and communities where, as a result of unstable states and vulnerable economies, pricing will be more profitable to waste management firms and brokers.

The fourth reason for the global waste trade is a racist and classist culture and ideology within northern communities and institutions that view toxic dumping on poor communities of color as perfectly acceptable. This ideology is best exemplified in an infamous internal World Bank memo authored in 1991 by Lawrence Summers, then chief economist and vice president of the World Bank:

> Shouldn't the World Bank be encouraging MORE migration of the dirty industries to the LDC [lesser developed countries]? I can think of three reasons. . . . 1) A given amount of health impairing pollution should be done in the country with the lowest cost, which will be the country with the lowest wages. I think the economic logic behind dumping a load of toxic waste in the lowest wage country is impeccable and we should face up to that. 2) I've always thought that under-populated countries in Africa are vastly UNDER-polluted, their air quality is probably vastly inefficiently low compared to Los Angeles or Mexico City. 3) The concern over an agent [pollutant] that causes a one in a million change in the odds of prostate cancer is obviously going to be much higher in a country where people [actually] survive to get prostate cancer than in a country [with higher mortality rates].[24]

No less disturbing than the content of the Summers memo are these points:

• In Summers's response to journalists and activists who later confronted him about the document he simply replied that the memo was meant to be "ironic." He did not deny the memo's content or its policy implications.

• The World Bank has indeed funded many toxic technology transfer schemes around the planet. Since the time of the Summers memo, those trends have continued, particularly in places like India and the Philippines, underscoring that the ideological position reflected in the memo was linked to the World Bank's actual policies.

• The consistency of the core reasoning of the memo with economic theory as it is taught to millions of university students each year and practiced by business leaders every day.[25] This is why global justice activists often critique the World Bank and IMF as sites of economic imperialism.[26]

• That the ideology that supports dumping on poor nations is also racist, because the peoples of most poor nations are primarily non-European peoples of color, and poverty is highly correlated with race around the globe.[27] Even the term *lesser developed country* (LDC) harkens back to theories of modernization that are infused with racism in that "economic development" is a code phrase for the degree to which a society can be considered civilized.

Despite the existence of numerous global conventions, international treaties, and national legislation in many countries that are intended to regulate and even prohibit the hazardous waste trade, toxic dumping in the global South continues. And environmental and social justice advocates continue to monitor and resist these practices.

The Waste Trade and Global Inequalities

Scholars of environmental justice studies and international relations have begun to tackle the question of global environmental inequality and racism. Much of the existing research on this topic comes from legal scholars wrestling with problems of international and domestic law on the waste trade—specifically, the legislation and treaties enacted to control these activities.[28] The legal literature centers mainly on one major pressing question: To what extent can domestic regulation and international agreements control or minimize the waste trade? A growing body of social science research has begun to pay attention to the social and economic driving forces behind the waste trade.[29] Even a cursory examination of the nations that are importing waste (legally or illegally) into their borders makes it clear that they are states on the geopolitical and economic periphery, have endured colonization in the past several centuries, and often are populated by a majority of people of color. For example, France colonized the African nation of Benin, which, even after independence, remains in debt to France and several international financial institutions as it attempts to rebuild its economy. French waste traders recently offered to pay Benin large sums of money as compensation for accepting toxic cargo. Pellow, Weinberg, and Schnaiberg reported in 2001 that Benin's motivation to accept such payment stemmed largely from its desire to repay its loans to France—hence, the term "toxic colonialism" and a brief explanation for one of the causes of global environmental inequality and racism.

Jennifer Clapp's book *Toxic Exports* is an outstanding analysis of the waste trade and the international NGO response to it through the Basel Convention, an international agreement among nations intended to ban waste shipments from global North to global South nations.[30] The history of the waste trade and NGO efforts to shape international policy that Clapp presents is authoritative. But there are questions that remain. For example, how is the global transfer of hazardous wastes and technologies linked to the struggle for environmental justice domestically in global North and South nations? How do transnational environmental justice movement networks mobilize against the hazardous waste trade outside official policy venues?

The movement against toxic dumping in poor neighborhoods and communities of color in the United States emerged during the 1980s, just as the movement against the global waste trade was taking shape. These two parallel events were not disconnected. Shortly after the movement for environmental justice in the United States made headlines in the early 1980s, activists and policymakers began to take notice of similar patterns of environmental inequality around the globe. The Basel Convention was signed in 1989, during the height of the environmental justice movement's visibility in the United States. It is also probably not coincidental that Greenpeace, which has been involved in struggles against environmental inequality across the United States, has been the principal advocate for a ban on the transnational trade in hazardous waste.[31]

Some scholars have noted that the majority of waste trading occurs among rich nation-states. For example, during the 1980s, at least 90 percent of the U.S. hazardous waste shipped abroad went to Canada, legally.[32] Much of the waste produced in North America and Europe is exchanged for compensation among these nations in a legalized system of toxic trade.[33] This is largely because the receiving nations have the technology to treat and manage such wastes and have negotiated for what they view as a fair price for the exchange. These practices began to change in the late 1980s as the cost of waste management skyrocketed, a result of relatively strict regulatory frameworks in northern nations. The export (whether through trade or dumping) of hazardous wastes to nations in the global South increased at the time and was problematic because, unlike trading among northern nations, few southern nations possess the infrastructure to properly treat and manage these wastes, and a fair price is hard to come by in such an unequal transaction. So while the majority of wastes produced in the North may remain within member nations of the Organization for Economic Cooperation and Development (OECD, and should, if the Basel Convention is adhered to), a significant portion of the most toxic of wastes still finds its way to the South.

African scholars and investigative journalists in global North nations were among the first and most outspoken critics of the international trade and dumping in hazardous wastes, defining this practice not only as an issue of ethics and morality but labeling it racist.[34] Philosopher Segun Gbadegesinn writes: "Since Africa has not been involved in generating the wastes, and since its people have not derived any comparable benefit from the outcome of the activities that led to these wastes (other than the accumulation of debt and poverty), it is unfair to impose on her the burden of waste disposal. . . . Toxic-waste dumping uses Africa as the dunghill for unwanted poisonous by-product of the excess consumption of developed nations."[35] Echoing this point of view, West African scholar Mutombo Mpanya writes,

> Africa is perceived of as a continent of immense jungles, populated by naïve people who are guided by corrupt and unintelligent leadership. . . . An official from Rodell Development, Inc., asked to comment on the possible health hazards its

toxic waste shipment, bound for Liberia, might pose to the indigenous population responded, "If anything happens to the Africans because of the waste, that's too bad. It's not our problem." Basically, the policies of industrial countries are designed to turn the lands of Africa and other Third World nations into landfills—the garbage dumps of prosperous industrial powers—in order to keep the Western world beautiful.[36]

Mpanya calls attention to the historic and contemporary potent and popular images of Africa as wild, untamed, corrupt and immoral, and unclean.[37] He also underscores the relational nature of environmental inequality and racism: that it is largely about concentrating hazards in others' backyards in order to keep one's own backyard clean. Another study reinforces this perspective: "Exporting hazardous waste results in higher environmental quality in the country of export, while the costs for proper waste management are externalized to the importing country."[38]

Government officials and activists in global South nations have been up in arms about the waste trade since the 1980s, and like some African scholars, they have emphasized the view that racism and historically rooted political economic relations are at the root of this practice. One West African head of state famously referred to waste dumping on the continent as "toxic colonialism."[39] An official of an overseas environmental organization told a U.S.-based journalist, "I am concerned that if U.S. people think of us as their backyard, they can also think of us as their outhouse."[40]

The waste trade really began when exporters targeted Africa in the 1980s, then moved to Latin America and South Asia, and then Eastern Europe, revealing both the power of transnational and grassroots movement networks to push traders from one part of the world to another and a racial global hierarchy that is all too familiar. For the past half-millennium, Africa has served as the world's primary colony for precious natural resources and slave labor. Viewed through this historical lens, the trajectory from slavery to colonialism and toxic waste dumping should surprise few observers. More generally, the dumping of toxic waste from global North to global South reflects the continuing corporate quest for the "path of least resistance"[41] as much as it embodies the practice of securing global race and class privileges.[42] And as R. Scott Frey argues, sending poisons to poor nations around the globe adheres to the historical pattern of siphoning wealth out of these former colonies, but it is also a new form of exploitation because it involves the export of "anti-wealth"[43]—the opposite of wealth—substances that drain a country's resources and poison its ability to produce resources in the future.

Race, class, and national inequalities are the primary drivers behind this drama, but the story is more complex. The environmental and environmental justice movements in the North have unwittingly contributed (at least partially) to the flow of destructive multinational corporate operations and hazardous wastes to the South.[44] As one environmental sociologist writes, "Ironically, the development of a North American environmental justice movement, which provided for greater environmental protection and greater citizen involvement in the [industrial facility and hazardous waste] permitting process, contributed to an intensified assault against native peoples in the Third World."[45]

Andrew Szasz documented the extraordinary success of the U.S. movement for ecopopulism during the 1980s and 1990s in its efforts to oppose the opening or expansion of landfills and incinerators across the nation at that time,[46] which built up enormous pressures on industry to find new dumping grounds. Grassroots opposition increased dramatically, and public hearings were a visible flash point for grievances. As one author noted: "Presently, it is very difficult in several industrialized countries to site new landfills or incinerators, and the situation has been described as an 'environmental emergency.' In fact, this is one of the reasons why hazardous wastes are being exported."[47]

Not surprisingly, the international dumping of hazardous wastes spiked during that period, creating a crisis of global environmental inequality and leading activists and governments to agree to the Basel Convention in 1989 and its amendment, the Basel Ban on North-South waste flows, in 1995.

There are two lessons here. The first is a cautionary tale in that social movements in the North should perhaps be more careful about how they approach the problem of domestic pollution, given the realities of economic globalization. There is a second, more hopeful, lesson: social movements have extraordinary power and can change the policies and practices of some of the world's largest corporations and most powerful governments. That nexus of state and corporate power is what I call the *political economic opportunity structure,* and movements are becoming adept at engaging those forces.[48] The real challenge is how to guide that power in ways that produce more progressive outcomes.

Environmental Justice Studies

Since the early 1970s, an increasing number of scholars in the United States have focused on the distributive impacts of environmental pollution on different social classes and racial and ethnic groups. Hundreds of studies have concluded that people of color and low-income populations bear a disproportionate burden of environmental exposure. Known variously as environmental racism, environmental inequality, or environmental injustice, this phenomenon has captured a great deal of scholarly attention in recent years.[49]

During this same period of scholarly interest in environmental inequality, a powerful social force, the environmental justice movement, emerged from within communities of color and poor and working-class white communities around the United States that have been inundated with air, water, and soil pollution.[50] The neighborhoods, playgrounds, schools, and workplaces where these populations "live, work, and play"[51] have been unequally burdened with a range of toxics, pollution, and hazardous and municipal waste from industry, agriculture, the military, and transportation sources.[52] The environmental justice movement is a grassroots response to the decline in quality of life as our society reinforces existing social—particularly racial, class, and gender—inequalities. As environmental degradation expands, we can expect that more and more communities will experience a similar outrage and contribute to the environmental justice movement.

Researchers from a range of disciplines conclude that the causes of environmental inequality and racism in the United States are varied and complex—for example:

- The tendency for corporations and governments to follow the path of least resistance in their decision making about where to locate toxic facilities and other environmental hazards.[53] Regulators and owners of noxious industries are very much aware that poor neighborhoods and communities of color have significantly less political clout than other groups, so there is less risk when they concentrate locally unwanted land uses (LULUs) in these areas.
- Housing market dynamics that frequently result in the colocation of people of color and environmental hazards.[54] Redlining and informal racist practices by lending institutions and real estate firms produce residential segregation and restrict the physical mobility of certain groups in or near toxic zones.[55]

• The exclusion of community voices and public participation from environmental policymaking processes, including urban planning and rural natural resource extractive activities, while special interests such as industry are often deeply involved.[56]

• The relative invisibility of people of color and working-class persons from the mainstream, national environmental movement in the United States.[57] This typically includes organizations like the Sierra Club, the Audubon Society, and the National Wildlife Federation. This absence of cultural and class diversity is believed to reflect a narrow worldview of environmental problems and solutions, which typically excludes the experiences of immigrants, poor people, indigenous peoples, and people of color.

• Racially and economically inequitable urban planning regimes and zoning practices.[58]

• The widespread violation of treaties with indigenous nations in North America.[59]

• A relatively weak labor and occupational health movement in the United States.[60]

One overarching social force that runs through each of these causes is institutional racism. Institutional racism is evident when institutions (governments, corporations, agencies, and even large environmental organizations) make decisions that appear to be race neutral in their intent but often result in racially unequal impacts.[61] The laissez-faire approach to zoning in the city of Houston (there is no zoning) is a case in point. Within such an arrangement, one would expect a matrix of factors to influence the location of LULUs, yet nearly all of that city's landfills are in communities of color, suggesting that race is the primary causal variable.[62] Thus, at the micro- and the macrosociological scales, environmental racism is linked not only to biased environmental policymaking but, more broadly, to racially biased practices within and across a myriad of institutions.

Class and gender inequalities are also deeply pronounced within environmental injustices. Class inequality is actually quite overt because market economies publicly embrace the ideology of wealth accumulation and profit for those who are able to achieve these goals over those who cannot. Thus, according to this logic, those who remain at or near the bottom of the economic pecking order—and therefore are more likely to live and work in environmentally hazardous conditions—are there because they simply have not availed themselves of what is available for the taking. Gender inequalities are integrally embedded in this system for four reasons. First, men tend to exercise the greatest control over states and corporations that produce environmental and economic inequalities, thus gaining the material and social benefits of both the economic and political power that results from and is reflected in environmental injustices. Second, men exercise the greatest control over national labor and mainstream environmental organizations combating economic and environmental inequalities and enjoy the status and credit for valiantly representing the interests of "the people" in national discourses and campaigns among such organizations seeking to combat the excesses of market economies.[63] Third, women tend to benefit the least from these struggles, as they are often physically and socially relegated to some of the most toxic residential and occupational spaces in communities and workplaces, and they are less politically visible because they tend to work for smaller, community-based, grassroots environmental justice and neighborhood organizations that rarely make headlines and survive on volunteer labor and small grants.[64] Finally, the very material landscapes being polluted and fought over in environmental justice struggles are deeply imbued with meanings that are raced, classed, and gendered and contained in local and global imaginaries, state policies, corporate practices, and activist resistance campaigns. The production of social inequalities by race, class, gender, and nation is not an aberration or the result of market failures. Rather, it is evidence of the normal, routine, functioning of capitalist economies. Modern market economies are *supposed* to produce social inequalities and environmental inequalities.[65]

The great majority of research in environmental justice studies is limited to the domestic sphere, particularly in the United States, so only recently have scholars begun to consider the fact that environmental inequality also occurs across nation-states or within other nations.[66] In this book, I explore how our understanding of environmental racism and inequality is transformed when we observe this phenomenon at work on a transnational scale in a global political economic system. What are the parallels and connections among corporate, nation-state, and social movement practices in the global North and those in the global South? Are the relationships between global North and global South environmental justice groups based on power sharing, consensus, and mutual respect, or do they reflect the inequalities and tensions we have observed domestically within the United States? What does all of this tell us about the ways economic globalization, racism, class inequalities, and environmental protection are changing in the twenty-first century? On a broader plane, how do environmental justice struggles reflect more fundamental problematics such as the tensions among capitalism, the state, the environment, and society in the context of late modernity? In the remainder of this chapter, I explore these questions, followed by a discussion of the methodological approach and an overview of the book.

Modernity, Environments, and Inequalities

Within environmental sociology, there are two broad schools of thought I address. The first is exemplified by the growing group of scholars writing on and advocating the idea of ecological modernization: the view that states and industries are improving their environmental performance with remarkable results that benefit the social and natural worlds. This school of thought is in keeping with more mainstream views of modernity, for example, where society is seen as evolving toward a state where free rational individuals are in control of their own affairs and those of the world. Modernity is a positive thing for the world, and "to be modern is to believe that the masterful transformation of the world is possible, indeed that it is likely."[67] The second school of environmental sociology I consider is characterized by scholars who view late modernity as a process that has created grave environmental and social problems around the globe. Within this school, I group together and consider the work of scholars of environmental justice studies, scholars advancing the treadmill of production theory, and those subscribing to the risk society thesis.

Ecological Modernization The core hypothesis of ecological modernization theory is that the design, performance, and evaluation of production processes have been increasingly based on ecological criteria rather than simply being rooted in a narrow economic calculus.[68] In contrast to other streams of environmental social science, and using an institutional analysis, ecological modernization theorists examine the extent to which, in the late twentieth and early twenty-first centuries, the environment has become an independent sphere in technology design, development, and decision making. These theorists argue that industrial society entered a new period in the 1980s, marked by new technologies, innovative entrepreneurs, and farsighted financiers who are bringing about a new generation of industrial innovation. This period, referred to as reconstruction, is marked by the emergence of an ecological sphere that exists independent of any other (economic, policy, or societal, for example).[69]

As a theory of industrial change, ecological modernization suggests that we have entered a new industrial revolution, one of restructuring production processes along ecological lines. But how does this approach locate and address the roots of the ecological crisis? Leading ecological modernization theorist Arthur Mol offers a perspective on this question: "Ecological modernization indicates the possibility of overcoming the environmental crisis while making use of the institutions of modernity and without leaving the path of modernization. The project aims at 'modernizing modernity' by repairing a structural design fault of modernity: the institutionalized destruction of nature."[70] In this way, Mol acknowledges that modernity appears to be predicated on environmental destruction, but only insofar as this is a design fault that needs repair. So in a problematic logical maneuver, ecological modernization maintains that both the cause of and solution to the environmental crisis lie within the structure of modernity itself.[71] While other scholars argue that capitalism and modernity are the roots of ecological harm and are therefore incompatible with sustainability, ecological modernization theorists[72] claim that economic development and rising environmental standards "go hand in hand."[73]

With regard to the question of transnational or global environmental trends, some ecological modernization scholars go so far as to argue that contrary to the "race to the bottom" or "pollution haven" thesis (wherein companies export environmental hazards to less economically powerful regions of the world), U.S. multinational chemical corporations are "exporting environmentalism" when they locate in global South nations like Brazil and Mexico and raise environmental standards in those nations.[74] Ecological modernization views economic growth as no longer necessarily linked to environmental harm.

Ecological modernization rests on at least two key problematic assumptions: that such technological improvements are economically feasible and that they are politically attainable. The growth and popularity of the ecological modernization thesis suggest several critical questions for analysis with respect to the transnational trade and dumping of hazardous wastes. First, is there sufficient evidence that the environment has become a key, independent factor in the technological design, development, and implementation of core waste-producing industries? If there indeed is evidence of progressive environmental change in these industries, what is the nature of the improvements, and why did the industry make these changes? In what ways is the ecological modernization (that is, the greening) of these industries linked to social movement action around environmental justice?

The evidence suggests that while some firms and states are incorporating ecological principles into their policies and practices, this is not nearly as widespread as ecological modernization proponents contend. Corporate-led globalization has continued to ravage the planet's fragile ecosystems, with few signs of abatement. Moreover, the social harms associated with late modern capitalism are producing continuing and growing social inequalities and political unrest.

The Treadmill of Production This model is a widely referenced framework emphasizing the origins of environmental problems in the political economy of advanced capitalist societies.[75] In a dramatic departure from the ecological modernization thesis (and indeed predating that school of thought by several years), Schnaiberg and others argue that capitalist economies behave like a "treadmill of production" that continuously creates ecological and social harm through a self-reinforcing mechanism of (generally) increasing rates of production and consumption. The root of the problem is the inherent need in market economies for capital investment in order to generate goods for sale on the market, income for workers, and legitimacy for nation-states. In other words, capitalism is a system that is ideologically wedded to infinite economic growth. However, there are severe social and ecological consequences. With regard to the ecosystem, capitalist market economies require increasing extraction of materials and energy from natural systems. When resources are limited, the treadmill searches for alternative sources rather than conserving and restructuring production. The treadmill operates in this way to maintain a positive and ever increasing rate of return on investments (although with routine fluctuations in economies, this is always variable). The state's role in this process is to facilitate capital growth and provide for social welfare and environmental protection, but these goals are dialectic: they exist in inherent tension.

The dialectic is reflective of two observations. First, most elements of ecological systems cannot fully meet both market value needs and social needs. And second, the treadmill of production prioritizes market value uses of ecosystems, despite the fact that other ecosystem uses are biological and social necessities for all classes of people. O'Connor reflects this point in his discussion of the "second contradiction" of capitalism, which involves capitalism's self-destructive tendency to appropriate a range of resources (labor power and natural resources, for instance) to the point at which the private costs of these activities spill over into the social arena.[76] The treadmill of production model reveals that, beginning in the post–World War II era—the era of late modern production—factories required greater material inputs than ever before as capitalism and consumer markets expanded nationally and internationally. Accordingly, this change necessitated the location, extraction, processing, and use of greater volumes of natural resources. Schnaiberg called these acts of natural resource depletion *withdrawals*. The other major change occurring in the late modern era was the exponential rise in the use of chemical inputs in production in the United States and other global North societies. Modernized factories were much more energy and chemical intensive in order to transform natural resources into market commodities more efficiently. This led to rising pollution levels, or what Schnaiberg called *additions*. As newer technologies were introduced over time, they were increasingly more chemical intensive or more reliant on automation and computerization, or both. So while creating more withdrawals and additions to and from the environment, these trends also led to the phaseout of many earlier forms of labor-intensive production, contributing to a massive disempowerment of labor. The environmental consequences of this arrangement include continued natural resource disruptions to feed the system, matched by increased pollution at the output end of the process. The social and economic impacts are also grave, because wealth is siphoned upward from the working classes to business and political elites, as wages at the bottom decrease, unemployment rises, and technology and automation displace even more workers to ensure cost savings and higher profits for industry and shareholders. Since these changes affect the more vulnerable segments of the working population, low-income persons, women, and people of color experience the impacts disproportionately. Thus, ecological disorganization and class, race, and gender inequalities are inherent by-products of the system.

Treadmill scholars view the relationships among the state, capital, residents, and workers over environmental protection as an "enduring conflict"[77] because the goals of profit, natural resource access, wage stability and job protection, public welfare, and environmental protection exist in tension. Thus, progressive environmental and social policies are likely to occur only as a result of massive disruptive action on the part of grassroots social movements. Moreover, the treadmill model implies that more democratic ownership and control over production and state functions could ameliorate social and ecological problems more than piecemeal policies aimed at reducing the use or volume of certain chemicals or efforts to control rates of consumption or consumer choice of certain products.

According to Schnaiberg et al., at the roots of these conflicts are power struggles over access to social, economic, and environmental resources, located primarily in class differences between the wealthy and the workers. As Schnaiberg and his collaborators demonstrate, the dynamics of the treadmill of production patterns hold true for environmental politics under globalization, as mainly northern elites and investors dominate the world economy and can shift much of the social and environmental costs of the treadmill to the South.[78]

The treadmill model presents a much more productive portrait of the relationship among capitalism, the state, the citizenry, and nature than does ecological modernization. Even so, it is fundamentally rooted in a Marxist orientation that pays less attention to the dynamics of racism and culture in the division of social and environmental benefits and costs. This study incorporates the treadmill model while avoiding what some critics might see as its heavy economic emphasis by paying closer attention to other social forces that drive and inform market economies.

The Risk Society A related theoretical framework is Ulrich Beck's "risk society" thesis.[79] As a number of scholars have noted, pollution, or industrial "smoke," was for much of the twentieth century viewed as "the smell of progress"[80] and was a strong indicator of economic vibrancy. When industrialists were challenged by neighborhood health activists or environmentalists concerned about ecological integrity or by workers concerned about occupational safety and health, their response has often been, "No smoke, no jobs,"[81] linking late modernity and subsistence to health-impairing and ecologically harmful practices. To be modern is to live in a risk society.

Modernity has become inextricably linked to the theory of the risk society. The risk society is marked not only by modern nation-state governance and citizenship practices, but also by a fundamental transformation in the relationship among capital, the state, and the environment—an exponential increase in the production and use of hazardous chemical substances. These practices emanate from the state and industry to civil society through consumption and disposal regimes, elevating the level of social and physical risk to scales never before imagined.[82] What this means is that the project of nation building, the very idea of the modern nation-state, is made possible by the existence of toxins—chemical poisons—that permeate every social institution, human body, and the natural world itself. To be modern, in short, is equated with a degree of manipulation of the natural and social worlds that puts both at great risk. To be modern also requires the subjugation and control over certain populations designated as others, less than fully deserving of citizenship, as a way of ameliorating the worst impacts of such a system on the privileged. These two tendencies are linked through the benefits that toxic systems of production produce for the privileged, and the externalization of the costs of that process to those spaces occupied by devalued and marginal others: people of color, the poor, indigenous persons, and even entire nations and regions of the globe.

Benton summarizes seven main features of Beck's risk society thesis.[83] He argues that according to Beck, the "new hazards" associated with the risk society:

- Are "unlimited in time and space"
- Are "socially unlimited in scope; potentially everyone is at risk"
- May be minimized but not eliminated
- Are irreversible
- Have "diverse sources, so that traditional methods of assigning responsibility do not work (Beck calls this 'organized nonliability')"

- Are "on such a scale or may be literally incalculable in ways that exceed the capacities of state or private organizations to provide insurance against them or compensation"
- May be identified and measured only by scientific means

In contrast to the ecological modernization thesis, the risk society model moves in quite the opposite direction. Ecological risks are deeply embedded in society and are ubiquitous and extremely harmful, yet frequently difficult to measure. Their existence and effects require expert knowledge, and even then, it is difficult to assign blame or develop policies that would address the problem since the sources of these risks are so diffuse. This is a problem for social movements in particular and for democratic governance in general.[84] However, as ubiquitous and diffuse as these toxics may be, Beck's view that locating sources of the problem is quite difficult is not always the case. Power is exercised by institutions that produce these toxins before they become diffuse, so if we can locate those institutional actors, pollution prevention is possible. As folksinger and activist Utah Phillips once stated, "The Earth is not dying—it is being killed. And the people who are killing it have names and addresses."[85]

At the root of the problem for Beck is a culture that places uncritical faith and acceptance in scientific rationality as a path to human improvement—one of the central tenets of European modernity. Science is a tool applied to the management of nature and people and is perhaps most effectively applied in industry and through markets. Hence, Beck views the power of private corporations as problematic in this model, because in most industrialized countries, they hold the greatest influence over research and development practices and scientific institutions. This produces a shift in power from the nation-state to corporations that enjoy hegemony over national and international scientific and political agendas.[86] The larger problem here is that unlike nation-states, private corporations do not operate on behalf of the citizenry and are not democratically run institutions. Thus, social change requires a different set of tools and strategies. In this regard, the risk society thesis shares common ground with the treadmill of production model. These two theories also emphasize the role of social inequality in this age of late modernity.

Beck points out that the politics of the distribution of environmental degradation favor more powerful communities over others:

> The history of risk distribution shows that, like wealth, risks adhere to the class pattern, only inversely; wealth accumulates at the top, risks at the bottom. . . . It is especially the cheaper residential areas for low-income groups near centers of industrial production that are permanently exposed to various pollutants in the air, the water and the soil. . . . Here it is not just this social filtering or amplification effect which produces class specific afflictions. The possibilities and abilities to deal with risk, avoid them or compensate for them are probably unequally divided among the various occupational and educational strata.[87]

Thus, advanced capitalism creates wealth for some and imposes risks on others, at least in the short term. In the long run, the problem of widespread global ecological harm, however, ends up returning to harm its creators in a boomerang effect. That is, the risks of modernity eventually haunt those who originally produced them. This generalization of risks unlimited in time or space is experienced by all persons, all groups, across the divides of social class and ethnicity.[88] Examples include the skin cancers associated with ozone depletion and the health problems that result from exposure to pesticide residues. In that sense, Beck acknowledges environmental inequality in the short term, while also maintaining a global, long-range view of what becomes a democratization of risk—thus, departing from the treadmill thesis.[89]

The risk society thesis puts forward the position that modernity is a fundamentally antiecological endeavor doomed to failure. The "design fault" that Mol views as easily fixable is, for Beck, the core of the problem and the death knell of society. The risk society thesis therefore has the real potential to mobilize all segments of society in favor of policies that would lead to improved environmental protection, if not sustainability. The politics of a risk society challenges the fundamental premises on which industrial society is constructed because it views modernity itself, and our most valued notions of civilization, progress, and development, as the root of the problem.[90]

The risk society thesis has much more in common with the treadmill of production model. However, the treadmill is more focused on the intersection of politics and markets than the role of science in this process. Furthermore, while the risk society approach argues forcefully that risk is ultimately universal (through the boomerang effect), the treadmill school views the problem as fundamentally about persistent social inequalities; therefore, as much as wealthy and elite populations may also experience ecological harm, their exposure pales by comparison to that of the working classes, and this is what keeps societies from coming together to address the problem. Both perspectives are useful for understanding the acute and widespread impacts of ecological harm. Ecological modernization is less useful precisely because it largely dismisses the intensity of social inequality and environmental degradation across societies.

In the remainder of the chapter, I consider trends and other evidence that speak to the direction in which states and market economies are moving global society with respect to environment and modernization.

Global Risk Society, Global Treadmills

Scholars from the risk society, treadmill of production, and environmental justice studies schools broadly agree that global volumes of pollution and toxics are not diminishing and that social inequality and industrial poisons have a curious habit of intermixing throughout the world. In this section, I examine some of the trends in global toxics production and environmental inequality.

Every living thing on the earth has been exposed to some level of human-made toxic substances. Lead, strontium-90, pesticides, and persistent organic pollutants (POPs) pervade our environment and reside in all of our bodies. This is a relatively new phenomenon, occurring mainly after World War II, as the production and use of hazardous substances increased exponentially in warfare, agriculture, electronics, and a range of industries, including transportation and housing. The considerable volume of hazardous wastes that were discovered in contaminated communities in the 1970s and 1980s in the United States were not anomalies; rather they were the by-product of a larger political economic reality that was ushered in during the post–World War II era. Since that time, the industrial and consumer economies have relied heavily on products made of chlorinated hydrocarbons. The size of these industries (plastics, oil, pharmaceuticals, and pesticides and chemicals) grew in response to increased demand from related industrial sectors and increased consumer demand for related products. The associated by-products were thus intensely toxic and increasingly ubiquitous (see figure 1.1). Historian Martin Melosi observes that the rise of urbanism and industrialization in the United States went hand in hand with pollution, in both practice and ideology.[91] So while the most egregious

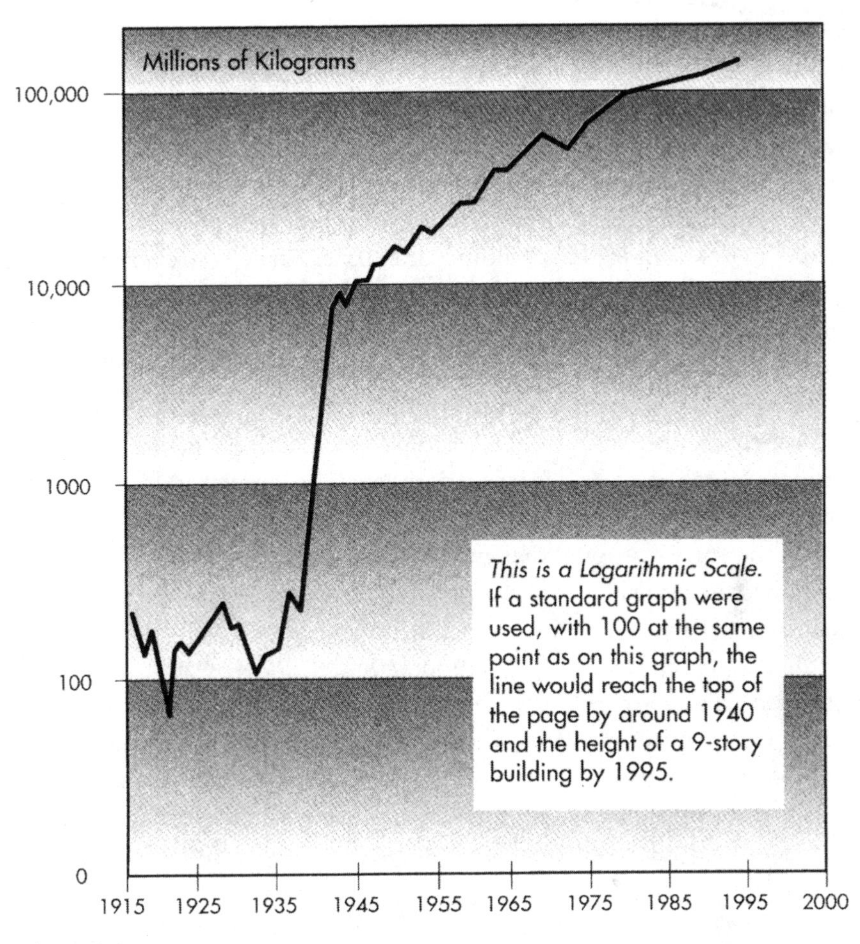

Figure 1.1
World production of synthetic organic chemicals. *Source: World Watch,* March/April 1997, p. 28.

manifestations of the risk society have emerged over the past half-century or so—the period often referred to as late modernity—this is an outcome with deep historical roots stemming from urbanism and industrialization that is centuries old. Today's hazardous wastes are the perilous physical and cultural residues of industrial production over the ages.

The Resource Conservation and Recovery Act (RCRA), a federal U.S. law, defines as "hazardous" those materials that may "pose a substantial present or potential hazard to human health or the environment when improperly treated, stored, transported, disposed of, or otherwise managed."[92] Although this definition is technically correct, the emphasis on the notion that such wastes present a danger only when improperly handled is severely misleading, since the very existence of these materials is hazardous. Moreover, the designation of materials as hazardous suggests that they lie at the extreme end of the production spectrum, when in fact they are at the core. The numerous industries that generate hazardous wastes are "the backbone of any industrial country, providing not only employment, but substantially contributing to the general welfare."[93] And as other nations move into the category of industrialized states, "hazardous waste has been an expected by-product of economic activity."[94] Hazardous wastes are generated by nearly every industry, and those industries that themselves generate few hazardous wastes nonetheless use products from hazardous waste–generating industries.[95] Societies in the global North are particularly ensconced in this process because they tend to be the largest producers of such toxics.

The role of science and technical knowledge of environmental risk is a curious one. On the one hand, with all the scientific evidence of the nature of risks to which we are exposed, one might think that any rational society would cease and desist in these activities so as to reduce the danger immediately. Yet we do not, precisely because we believe we can manage these risks ourselves and still have all of the conveniences associated with late modern capitalism. Given this orientation, it is quite surprising to consider the absence of rigorous, longitudinal, and definitive data on the health and environmental risks of our chemical-intensive lifestyle. Despite Rachel Carson's own research and dire warnings in her classic book *Silent Spring,* we continue to produce and use even more chemicals and have taken few steps to understand their potential impacts before doing so. This is why many scientists, policymakers, and environmental activists are calling for the adoption of the precautionary principle.[96]

The precautionary principle takes the position that if there is reasonable indication that a chemical may be unsafe, we should refrain from using it, even if there is not yet conclusive scientific evidence to that effect. This is a regulatory approach that shifts the burden of proof that chemicals are safe onto the producers rather than allowing them to essentially test these materials on an unwitting public. The current regulatory framework in the United States presumes chemicals are "innocent" until proven "guilty" and simply releases them into widespread use until there is reason to believe they are unsafe. The consequences have been disastrous. We have scarcely any toxicological data on the more than 80,000 chemicals in use today. The extent of the production of toxins and their associated risk in the United States is staggering. The United States produces nearly 6 trillion pounds of chemicals annually.[97] Toxic materials exposure can cause genetic defects, reproductive disorders, cancers, neurological damage, and the destruction of immune systems. "Wherever there is industry, there are hazardous wastes."[98]

The evidence of risk and disease associated with industrialization is mounting. In February 2004, scientists with the USEPA estimated that one in six pregnant women in the United States has enough mercury in her blood to pose a risk of brain damage to her developing child. This new estimate is double that of a previous calculation.[99] Mercury is a heavy metal, and when it is ingested or spilled in the environment in tiny amounts, it can wreak havoc on the nervous system of humans and other living beings. Effects include damage to the brain, lung, and kidneys, and death. Mercury is released into the environment primarily by power plants and waste incinerators polluting the air and is then deposited into oceans and other waterways, where humans and other animals ingest it and it bioaccumulates throughout the food chain.

The U.S. Centers for Disease Control and Prevention (CDC) released a study in January 2003 in which researchers tested a sample of more than 9000 individuals across the United States. They found pesticides in 100 percent of their bodies.[100]

Polybrominated diphenyl ethers (PBDEs) are a little-known class of neurotoxic chemicals found in computers, televisions, cars, furniture, and other common products that global North consumers use every day. They are ubiquitous not only because they are contained in so many consumer products but because they also leak into the environment during production, use, and disposal. As a result, they are found in household dust, indoor and outdoor air, watersheds, and the body tissues of dozens of animal species around the world, including humans. Women's breast milk in the United States, Europe, and Canada contains high levels of PBDE, and most residents in the United States are believed to carry this chemical in their bodies at unsafe levels.[101]

Despite the more influential environmental and labor movement community in Europe, European Union (EU) nations continue to pollute at an alarming pace as well. One in five persons employed in EU nations is exposed to carcinogenic agents on the job. Cancer, asthma, and neuropsychiatric disorders are some of the illnesses associated with the 100,000 chemicals and biological agents marketed in the EU, according to the European Agency for Safety and Health at Work. Approximately two-thirds of the 30,000 most commonly used chemicals in the EU have not been fully tested for their potential health impacts on humans or the environment. Chemicals introduced since 1981 undergo such tests, but the older ones remain untested.[102] These facts speak to Beck's contention that many of these risks haunt us, yet they are mysterious and largely unknown. The question of intergenerational impacts emerges here as well, because we have a greater potential to harm future generations (irreparably) than any other previous one.

Exported waste may eventually come back to haunt us in the United States and other global North nations that export it so freely. "It's possible that we could send sludge to the Caribbean and they might use it on spinach or other vegetables [that we may later import]," noted Wendy Greider, an official at the USEPA's Office of International Activities.[103] And since the Food and Drug Administration checks only a small portion of food entering the United States, hazardous wastes that were exported abroad could easily end up on the dinner table. In addition to agricultural pesticide life cycles, air and water pollution in the South knows no boundaries and easily loops back to harm residents and consumers in northern nations. This is what some scholars have called a "boomerang effect" or the "circle of poison."[104] In later chapters, we consider the practice whereby activists in global South nations send toxic waste back to the original exporting nation (in the North), a more direct version of the boomerang effect and perhaps one of the more poetic and symbolic forms of environmental justice NGOs have devised in recent years, called "Return to Sender."

From a sociological perspective we can conclude that there are social or cultural reasons driving behaviors and trends of a risk society. These practices are facilitated and reinforced by powerful institutions and by consumers and workers who have grown dependent on toxic systems of production. In the global North, we recognize that chemicals are indeed hazardous to life itself, yet we adopt the belief that if we use them responsibly, they can produce collective benefits. Consider this news headline: "Federal Judge Rules Chemicals Used in Executions Are Humane."[105] In this sense, chemical and biological hazards are viewed not only as the hallmark of a nation's embrace of modernity, but also as a marker of humanity. After all, we have standards of ethical behavior and codes of conduct. Consider another headline: "Study Clears Pesticide Tests with Humans."[106] Lest we worry that this revelation is indicative of irresponsible use of otherwise hazardous substances, the article notes that the USEPA will be "allowed to use data from studies in which humans are intentionally doused with pesticides and other toxic substances, as long as strict scientific and ethical standards are met, a National Academy of Sciences report has concluded."[107]

All of this behavior has real consequences. The United Nations' Millennium Environmental Assessment reported in no uncertain terms that the global environmental crisis is dire and worsening by the year.[108] The news is not all bad, however, particularly in the EU, where environmental organizations have successfully pressured states and industries to pass progressive legislation that would mandate the risk evaluation of chemical substances (such as the REACH policy, that is, the Registration Evaluation Authorization of Chemicals) and the requirement that electronics manufacturers produce their goods with fewer toxic substances and take back those products for recycling at the end of their consumer life.[109] These laws and policies may lead to environmental improvements in Europe. The real question is whether they will do so by encouraging a shift of toxics southward. As internationally acclaimed Filipino environmental justice activist Von Hernandez put it, "In Europe right now there is the REACH directive, an EU-wide policy on regulating chemicals. So the debate in Europe as far as chemicals are concerned is right now more advanced than the rest of the world. The rest of the world has to catch up . . . because developments in Europe will impact chemical production in the US, in Japan, in Asia. Because otherwise you would see a situation where discredited chemical manufacturing or chemicals being produced in Europe would be moving South again, similar to the developments we've seen with the adoption of the Basel Convention."[110]

Global Inequality Trends: A Treadmill of Environmental Injustice

The roots of global environmental injustice lie mainly in the production and consumption patterns of northern societies, which have unequal impacts on the poor and people of color worldwide.

The *State of the World 2004* report concluded that North America and Western Europe, representing just 12 percent of the world's population, account for fully 60 percent of the consumption of the world's natural resources. By contrast, the one-third of the world's population living in South Asia and sub-Saharan Africa accounts for only 3.2 percent of this consumption.[111]

The world's richest nations are depleting natural resources at an unprecedented rate. The concept of an ecological footprint is intended to capture the extent to which a nation can support its resource consumption with its own available ecological capacity. In 2000, the United States was the nation with the largest per capita ecological footprint on the globe, with a footprint of 23.7 acres per capita. A sustainable footprint for the United States would be 4.6 acres.[112] According to a United Nations–sponsored study released in 2002, citizens in the United States and Canada may enjoy a cleaner environment "at the expense of global natural resources and climate."[113] The report states, "Each Canadian and American consumes nine times more gasoline than any other person in the world. With only about 5 percent of the world's population, both countries accounted for 25.8 percent of global emissions of the major greenhouse gas carbon dioxide, created by the combustion of coal, oil, and gas."[114]

Affluence among nations is highly correlated with environmental harm. In the early 1990s, the twenty-four richest and most heavily industrialized nations collectively produced 98 percent of all hazardous wastes.[115] Rich nations in general are not reducing the level of hazardous wastes produced today: "OECD countries presently create 220 pounds of legally-hazardous waste per person per year. By 2020, per-capita production will rise 47% to 320 pounds per person per year and, because of growing population, total OECD hazardous waste will increase 60% to 194 million tons each year. All of this will eventually enter the general environment and significant portions of it will enter food chains."[116]

Given the high level of toxicity of everyday life in the global North, if states and corporations are not planning to reduce toxic inputs into production, then it makes sense to seek outlets for dumping some of the most hazardous substances elsewhere, to reduce exposure to these dangers. A logical approach would be to export these wastes to global South communities, which may allow us to embrace the idea of ecological modernization because the more visible dimensions of pollution are now "out of sight, out of mind" (which also occurs as a result of domestic environmental inequality and racism).[117]

The classic report of the World Commission on Environment and Development, *Our Common Future,* stated clearly: "Most hazardous waste is generated in industrial countries. However, exporting waste results in potential risks primarily to people in importing countries, who do not share in the benefits of the waste generating production processes. The people who share the potential risks have little, if any, practical influence on the decision to import these wastes."[118] The export of hazardous waste and materials to nations with less stringent environmental standards is not only an example of environmental inequality and racism. It is also a clear violation of the United Nations' Stockholm Declaration, which states in Principle 21, "States have, in accordance with the Charter of the United Nations and the principles of international law . . . the responsibility to ensure that activities within their jurisdiction or control do not cause damage to the environment of other States or of areas beyond the limits of national jurisdiction."[119]

The evidence from scientific, social scientific, and governmental studies indicates quite strongly that social and environmental inequalities persist within and between nations, lending support to the treadmill of production, risk society, and environmental inequality and racism theses. Toxics abound globally and are present in all nations and all living beings. The contention by ecological modernization theorists that environmental conditions in the world are improving is both supported and contradicted by the evidence. The model is supported when we observe environmental improvements taking hold in global North nations (such as EU-wide legislation forcing industry to design electronics components for end-of-life recycling). However, ecological modernization is contradicted because such improvements in the North may largely be due to the actions of corporations that shift many of the most toxic industrial hazards southward, producing environmental inequalities. One's view of ecological modernization therefore depends on what spatial scale one considers and how far along the commodity chain one follows a product. Integrating an environmental justice analysis into the ecological modernization framework is useful because there may indeed be environmental pollution improvements in the North, but they are often only in certain privileged communities (wealthy and white), while others (people of color and poor populations) see an intensification of environmental hazards. Globally, the same observations hold true: when the United States, Canada, or the United Kingdom has a national improvement in environmental indicators, it may often be because these hazards have simply been shifted geographically southward.[120] Thus, ecological modernization is possible precisely as a result of global environmental inequality and racism.

Methodological Approach

I gathered data for this study between 1998 and 2006 using four principal research methods. First, I conducted a review of the literature on the transnational waste trade, environmental justice studies, environmental sociology, social movements, and race theory. Based on this review, it was clear that the four interventions identified in this chapter have not been sufficiently addressed by scholars. Second, from several libraries and archives, I conducted content analyses of newspaper articles, government documents, NGO reports, and books on global environmental policy conflicts and transnational movement networks in the United States, the Caribbean, Asia, Africa, and Europe between 1987 and 2006, a period that marks the beginning of the era of transnational waste trade. Sources for these data include hundreds of memos, reports, internal documents, and studies from various grassroots and advocacy organizations.[121] Third, I conducted semi-structured interviews with three dozen leading international environmental justice activists from around the world where they and their networks have been caught up in key struggles that have defined the politics of the global waste trade and the global environmental justice movement. Fourth, I attended, organized, or participated in a number of national and international conferences focusing on global environmental justice and human rights, which allowed me to gain access to additional documents, reports, and studies on this topic.[122]

This book is part of a broader program of advocacy research: the effort by scholars to produce research that is accessible to and in the service of the people we write about, as well as the general public. Although it is enjoying a revival,[123] there is a long tradition of this kind of research in the social sciences. For example, in their book *Liberation Sociology,* Feagin and Vera examine the largely unacknowledged history of scholars working on advocacy research efforts in vulnerable communities in the United States and around the world.[124] More than 150 years ago, sociologist Karl Marx wrote, "Philosophers have only interpreted the world, in various ways; the point, however, is to change it."[125] Activist-scholars like W. E. B. DuBois continued this project by authoring outstanding works of critical scholarship on racial inequality in the United States and around the globe while advocating and making social change.[126] Building on this history of liberation sociology, Feagin and Vera contend that the "ultimate test of social science is not some type of propositional theory building but whether it sharpens our understanding and helps to build more just and democratic societies."[127] I concur. Some scholars argue that social scientists in particular have an obligation to engage the world and offer our analytical skills with the aim of improving society. As Philo and Miller write, "A large part of humanity is being obliterated by the social, material and cultural relationships which form our world. It can be painful and perhaps professionally damaging to look at such issues and to ask critical questions about social outcomes and power. . . . But for academics to look away from the forces which limit and damage the lives of so many, gives at best an inadequate social science and at worst is an intellectual treason—just fiddling while the world burns."[128]

I call what I do *critical advocacy research:* I participate in social change efforts while also stepping back and employing a reflexive analysis of that work. It is my conviction that social scientists can be part of a major movement that can achieve social change. Through my work with several NGOs and foundations,[129] my research and ideas on global environmental justice struggles developed a great deal, and I hope this book illuminates more productive ways to think about and orient action against the problem of global toxics and environmental inequality.

Overview of the Book

Chapter 2 examines the emergence of political, economic, and ideological forces in world history that produced controlling discourses and practices concerning racial difference and the natural environment. I consider the ways in which resistance to this system of domination of people and ecosystems can be integrated with—and used to extend—social movement theory. Chapter 3 charts out a portrait of some of the major transnational environmental justice movement networks operating around the world today. I present some of the core concerns, controversies, and strategies these activists, organizations, and networks address in this era of globalization. I begin chapter 4 by examining the seminal waste issue in domestic and transnational environmental conflicts: garbage. I then move, in chapter 5, to consider the legacy of the Green Revolution and international efforts to bring agricultural "development" to the global South through the transfer of countless tons of toxic pesticides from the North. Chapter 6 examines the latest scourge of transnational environmental inequality: the dumping and remanufacturing of high-tech and electronics products (e-waste) in the South. The journey mapped here moves from the crudest and age-old dumping practices—garbage—to what, for some, exemplifies the postmodern condition: high-technology products that allow for the compression of space and time, and the sharing and reproduction of cultures across national borders in ways that our ancestors could only have dreamed of. Unfortunately, postmodernity and global cultures imprison us more than they liberate us from either our earthly origins and limitations or our tendencies to create and struggle with hierarchies and inequalities. My goal in chapter 7—and throughout the rest of the book—is to contribute to debates and actions that will move us more productively along paths toward environmental justice, human rights, and sustainability.

2

Race, Class, Environment, and Resistance

In this chapter, I build on the work of scholars examining the connections between the emergence of nation-states, capitalism, and the ideological foundations supporting these structures in world history, and systems of control over nature, people of color, and working-class populations. I then engage the writings of social movement theorists studying transnational protest behavior and offer critical points of intervention for new directions in that field.

Race, Modernity, and the Environment

Contemporary racism by global North nations builds on a rich tradition of conquest, slavery, and colonization. Racism has been an organizing principle of the modern world system since the rise of European states centuries ago.[1] The rise of capitalism and European nation-states was facilitated through intersections between the ideological and structural domination of nature and non-European peoples. Modernity, then, is a global racial project.[2] Historian Kirkpatrick Sale maintains that capitalism took hold with such strength in sixteenth-century Europe because, after centuries of death, disease, violence, and misery associated with feudalism, the Black Plague, and a series of wars, Europeans were eager to embrace a much more material reality—one that could be rationally seen, touched, and measured scientifically. This was a core cultural basis for capitalism and modernity. But this emergent sentiment was underscored by a haunting fear of, and desire to dominate, nature. And because non-European peoples came to be associated with nature, they too were to be dominated. Enrique Dussel contends that both modernity and capitalism depend on the destruction of nature while relying on its wealth to fuel production. This system depends on the labor of poor people and various ethnic groups, while systematically disempowering and subjugating those populations.[3]

A number of historians and philosophers agree that a core part of the modernity project has been the separation of "civilized" (human) culture from nature.[4] That is, through what Rousseau called the social contract, "subpersons" and "savages" were excluded from the polity since they were equated with nature. Thus, nature and the spaces certain people (or "subpersons") occupied were devalued, or perhaps selectively valued insofar as they provided a strategic benefit to the modernity project. Thus, the meaning of space and race codeveloped to produce racism and environmental harm as one integrated process as European expansion took hold during the sixteenth century. In other words, the ideological, cultural, psychological, and physical harm visited on people of color was supported and made possible by a system that did the same to nature. As Joel Kovel writes: "The domination of nature, combined with more straightforward political, cultural, and economic patterns, is what gives our racism its unique character: it is what makes white racism white. . . . The dominant cultural sweep of the whitening of racism is there for all to see as a species of the domination of nature."[5]

This arrangement created a global racial and ecological reality. W. E. B. DuBois described this process when recognizing the links between the oppression of people of color and the manipulation of nature in a global capitalist system:

> That dark and vast sea of human labor in China and India, the South Seas and all Africa: in the West Indies and Central America and in the United States—that great majority of mankind, on whose bent and broken backs rest today the founding stones of modern industry—shares a common destiny; it is despised and rejected by race and color; paid a wage below the level of decent living; driven, beaten, prisoned, and enslaved in all but name; spawning the world's raw material and luxury—cotton, wool, coffee, tea, cocoa, palm oil, fibers, spices, rubber, silks, lumber, copper, gold, diamonds, leather—how shall we end the list and where? All these are gathered up at prices lowest of the low, manufactured, transformed, and transported at fabulous gain; and the resultant wealth is distributed and displayed and made the basis of world power and universal dominion and armed arrogance in London and Paris, Berlin and Rome, New York and Rio de Janeiro.[6]

In relational fashion, the wealth of nature and the labor power of subjugated people were sapped for the benefit of a privileged few. People, or "others," on the lower rungs within this social hierarchy are generally perceived as closer to nature because of their allegedly primitive, land-based cultures, and this ideology operates in many forms today. This form of racialization functions to diminish the rights of people of color, indigenous peoples, and immigrants based on the racial categories they occupy. This racist logic, which gives ethnic minorities an "animalistic" quality, justifies the concentration of people of color in jobs and residential spaces that are particularly dirty or hazardous.

Scholars including Neil Smith and Joel Kovel have demonstrated that through the psychology of whiteness, many European Americans view the inner cities where many African Americans are concentrated as wild places, or a kind of frontier populated by shadowy, dark, subhuman others—places to be feared, avoided, or conquered.[7] Linguist Otto Santa Ana analyzed media metaphors used to describe Mexican immigration into the United States and found that naturalistic images like "waves on a beach," "parasites," "weeds," and "animals" flowing into our otherwise ethnically pure and homogeneous nation are frequently prevalent.[8] In a post-9/11 era, where the term *terrorism* conjures up brown faces (this time, Arab and South Asian, not Latino) driven by irrational impulses to wreak havoc on our society, we see stickers appearing on cars throughout the United States that read "Terrorist Hunting Permit" (issued in all fifty states). This ideological discourse serves to present people of color as in need of constant discipline and order lest they return to their "wild" ways. Those on the higher rungs of this racial hierarchy are left with the "responsibility" of providing the discipline for those who do not know better. This is a contemporary manifestation of Rudyard Kipling's "The White Man's Burden" and can also be read into many statements that George W. Bush made before, during, and after the 2003 invasion of Iraq regarding the U.S. military's intention of "liberating" that nation and bringing democracy to a people who had supposedly never known such a political tradition. The name of the military campaign itself, Operation Iraqi Freedom, embodies this message.

Globalized Race and Class Hierarchies, Privatized Inequalities

Racism is a global phenomenon. Scholars of globalization and race tend to fall into one of two schools. One school views globalization as a process that is liberating because it depoliticizes race through the intensified flow of people, goods, ideas, and capital across national borders, all of which allow us to escape the restrictive confines of racial identities imposed on us by nation-states.[9] Technological advances like jumbo jet airplanes, the Internet, fax machines, videoconferencing, and cell phones compress time and space. They also create new ethnoscapes and global cultures that transcend geography as music, language, religion, images, and voices flow around the globe, shaping consciousness, politics, economies, and ecosystems in new ways.

Another school of globalization and race argues that although these rapid cultural transformations are indeed taking place, they are reflective of both old and new hegemonies that have maintained the subjugated position of indigenous, poor, and ethnic populations around the globe much more than they have elevated them. For example, studies reveal that the electronics infrastructure that makes possible much of these 'liberating' global cultural changes comes at the expense of devastated ecosystems from which the raw materials are extracted and harms the health of workers who manufacture and recycle such products at every stage of the commodity chain.[10] The economic and political benefits of these arrangements are siphoned upward and enjoyed by a small minority of wealthy investors and firm owners. Similar social and ecological dynamics are widely documented in the energy sector, which is at the core of the global economy, making possible its very existence.[11] Moreover, despite all of the talk about globalization, the continued use of nationalism and nation-state borders to reinforce ethnic, religious, and racial hierarchies, and exclusions challenges the notion that cultural fluidity is the order of the day. With recent events surrounding the emergence of the war on terror and related racial projects, such as xenophobic immigration policies in the United States and the European Union, the links between race and nation remain quite strong in both discourse and policy.[12]

The globalization of racism is nothing new. It has been an axis of political, cultural, and economic organization for international capitalist enterprises for centuries.[13] This is a critical point, because much of the most widely cited literature on the ill effects of economic globalization rarely characterizes the inequalities and injustices that accompany this process as racialized.[14] Moreover, as much as this was an extension of empire and influence by nation-states, market forces have been at the core of globalization since the beginning.

Linking Class and Race through Markets and Politics

Barlow argues that one of the central problems for racial justice movements is the increasing inability of states to regulate markets and highly mobile capital networks in a global economy. I concur and argue that theorists studying racism will have a limited understanding of racial inequality if the primary conceptualization of racism's roots is in either state formation or capitalism, because the answer lies within both spheres. Thus, as I discuss later in this chapter, movements for racial and environmental justice must successfully engage both states and markets—what I term political economic opportunity structures.[15] The political economic opportunity structure must be confronted, or social inequalities will continue to expand. While cultural theorists celebrate the alleged leveling of social differences owing to globalization, actual class and racial inequalities around the globe and between North and South have only worsened in the past half-century—the same period during which the late modern stage of capitalism took hold. The class inequalities alone are extraordinary:

> As a result of capital's greater leverage, one of the most pronounced effects of market globalization is the rapid growth of income and wealth inequality, both within nations and globally. That is, the rich are getting richer and the poor are getting poorer at a rate never before seen in world history. . . . As a result, the global gap between the richest and poorest people has rapidly widened. In 1800, the ratio of the per capita income of the richest fifth to the poorest fifth of the world's population was two to one. By 1945, the ratio was twenty to one. In 1975, the ratio had risen to forty to one. The assets of the world's 358 richest individuals now exceed the combined incomes of 45% of the world's population. A similar trend toward growing inequality is taking places within the MDCs [most developed countries] themselves.[16]

The income gap between the fifth of the world's people living in the richest countries and the fifth living in the poorest was seventy-four to one in 1997, up from sixty to one in 1990 and thirty to one in 1960. This is evidence that class inequalities are not only extant but growing over time. By 1997, the richest 20 percent of the world's population captured 86 percent of world income, with the poorest 20 percent holding onto a mere 1 percent.[17] The gap between rich and poor between and within nations is on the rise.

Other studies confirm these unsettling trends:

> The median income of the tenth of the world's population living in the richest countries was 77 times greater than the tenth in the poorest in 1980, but by 1999, the richest earned 122 times the poorest countries. Entire regions of the developing world are falling alarmingly behind the wealthiest countries that compose the Organization for Economic Cooperation and Development (OECD). Sub-Saharan Africa's per capita income was one sixth of OECD countries in 1975, but fell to one fourteenth of OECD per capita income in 2000. Over the same period, Latin America and the Caribbean per capita income fell from less than half that of OECD countries to less than a third, and Arab countries' per capita income fell from one quarter of that of OECD countries to one fifth.[18]

Wealth and income continue to be thrust upward within and between nations, while the poor work harder for less remuneration and the rich are richer than ever before in many countries. The assumption that material conditions will improve for the world's poor through economic globalization is patently false and historically counterfactual. The wealth and income gaps between countries have grown geometrically over the past century.[19] These global inequalities by nation and class—which world systems theorists have so thoroughly studied—are also highly correlated with global racial inequalities: "Pick any relevant sociological indicator—life expectancy, infant mortality, literacy, access to health care, income level—and apply it in virtually any setting, global, regional, or local, and the results will be the same: the worldwide correlation of wealth and well-being with white skin and European descent, and of poverty and immiseration with dark skin and 'otherness.' [There is a] . . . planetary correlation of darkness and poverty."[20]

It is important to note the power of capital in this process, particularly transnational corporations, and the ideological power of privatization in relation to race. Many scholars maintain that the face of racism has changed significantly in the United States and globally over the past several decades (beginning as early as World War II); today it is less overt, often invisible, and embedded in everyday institutional practices.[21] Since racism remains primarily a structural and material relationship, what may have changed most in recent years is the nature of the discourse and rhetoric around difference and (in)equality. Howard Winant describes the change in racial common sense as a "historical break," where race relations were transformed globally.[22] The time was around 1945 at the dawn of the post–World War II era, as independence movements in Africa, Asia, and Latin America were gaining strength and challenging global white supremacy (at least as promoted by northern nation states). Independence and decolonization movements in the global South, combined with the civil rights movement in the United States (a segment of the "South of the North"), changed the terrain of acceptable racial behaviors and discourses, moving them from their previously more overt and blatant forms to more covert and less visible practices. What is missing from this story, however, is the fact that at that same moment, we saw, first, the intensified growth of economic globalization, as a result of the Bretton Woods Agreement, which produced the International Monetary Fund (IMF) and the World Bank, facilitating the use of market expansion as a way to extend and maintain U.S. and European control over global South nations through the increased prominence of transnational corporations (TNCs) (that is, neocolonialism). We also saw the exponential rise in the manufacture and use of industrial chemicals for inputs into a host of production processes, especially agriculture.[23] Both changes had enormous implications because together, they seriously undermined the gains of the civil rights and labor movements and the general antiracist struggle in the United States and undercut the global benefits of the anticolonial struggles occurring throughout the global South. These two developments are intimately linked because the TNCs themselves were the ones creating and pushing both globalization and toxins on world markets, facilitating greater control over nations, communities, human bodies, and the natural world itself. However, racial theorists tend not to address this story and instead focus on the activities of nation-states as primary drivers of racial formation (while often paying less attention to capital) and also tend to overlook the devastation brought on by the antienvironmental practices and human rights violations that intensified at that historical moment. For example, Monsanto Corporation entered India in 1949, two years after that nation earned its hard-won independence. India has been fighting against this corporation's pollution and its stranglehold over its food systems ever since. Although scholars are fond of describing India as a postcolonial state, they should not forget that it is still battling corporate neocolonialism.

In keeping with Winant's concept of the historic break in race relations, it was no longer acceptable for nation-states to overtly or directly dominate other (southern) nations, so their corporations stepped up to play that role, under the guise of market liberalization. Thus, the rise of TNCs in the post–World War II era was, to a great extent, a response to global grassroots independence movements—a reassertion of global racial and class dominance.[24] Many activists and scholars who focus on the power of TNCs' view these organizations simply as a global extension of a Marxist class-focused inequality analysis, but they forget that corporations are just as race conscious and just as complicit in producing and maintaining racial hierarchies as nation-states.[25] While race theorists concede that nations constitute "racial states" in that they reproduce racial logics, the corporation as a racial institution is a concept in need of development and a reality that must be confronted.[26] That is, during the same period when the historic break in race relations occurred, the rise of corporate power signaled another break wherein states no longer held the monopoly (or even hegemony) on defining and producing racial hierarchies. Corporations, particularly TNCs, now have as much influence over shaping race relations and the physical terrain on which these relations operate as nation-states ever did. Carl Boggs calls this process "corporate colonization," which has produced a "corporate polity" in which these private organizations have as much political and cultural power as governments.[27] Corporate colonization is fueled in part by racial ideologies.

Racism and Class Domination as Toxicity: The Ecology of Social Inequality

For well over a century, the use of words like *poison* and *toxic* has been common in writings about race and racism by activists, novelists, and scholars. This is a powerful and productive way to capture the harm that racism does to both its victims and its perpetrators or beneficiaries. Cornel West describes the paradox of racism in the United States as a poison that reveals deep contradictions and tensions in this nation, which have periodically erupted in violence, revolts, and wars over the years.[28] Lani Guinier and Gerald Torres make use of this terminology as well in their book *The Miner's Canary*. They write:

> Race, for us, is like the miner's canary. Miners often carried a canary into the mine alongside them. The canary's more fragile respiratory system would cause it to collapse from *noxious gases* long before humans were affected, thus alerting the miners to danger. The canary's distress signaled that it was time to get out of the mine because the air was becoming too *poisonous* to breathe. Those who are racially marginalized are like the miner's canary: their distress is the first sign of a danger that *threatens us all.* It is easy enough to think that when we sacrifice this canary, the only harm is to communities of color. Yet others ignore problems that converge around racial minorities at their own peril, for these problems are symptoms warning us that we are all at risk.[29]

Guinier and Torres also address the pervasive idea of color blindness and link it to the "poison" metaphor of racism, through a new concept they term "political race":

> Political race as a concept encompasses the view that race still matters because racialized communities provide the early warning signs of *poison* in the social atmosphere. And then it encourages us to do something different from what has been done in the past with that understanding. Political race tells us *that we need to change the air* in the mines. If you care to look, you can see the canary alerting us to both danger and promise. The project of political race challenges both those on the right who say race is not real as well as those on the left who say it is real but we cannot talk about it.[30]

Guinier and Torres's concept of political race pushes us to challenge the toxicity of racism because it "threatens us all," not just people of color who may be its most direct targets. West, Guinier, and Torres were not the first to use these environmental and chemical metaphors for describing racism. In 1881, the great human rights activist Frederick Douglass wrote about racism in the following way: "In nearly every department of American life [black Americans] are confronted by this insidious influence. *It fills the air.*"[31] Martin Luther King Jr. once declared, "White America is still *poisoned* by racism, which is as native to our soil as pine trees, sagebrush and buffalo grass."[32] King pointedly contended that racism is a poison but that it also harms white America as it simultaneously devastates communities of color.

Guinier and Torres were inspired to use these environmental metaphors by a law review article published in 1953 on the erosion of Indian rights in the United States. In that article, legendary Indian rights attorney Felix Cohen wrote, "Like the miner's canary, the Indian marks the shift from fresh air to poison gas in our political atmosphere, and our treatment of the Indian . . . marks the rise and fall in our democratic faith." Crime mystery novelist Walter Mosley, best known for his writings about the experiences of African

Americans in post–World War II Los Angeles struggling against discrimination, wrote in his celebrated book *White Butterfly* "The air we breathed was racist." Each of these scholars, writers, and activists maintains that through webs of social, cultural, political, and economic interdependence, the violence and harm of racism poison all of society; no one is immune.[33]

These theorists and activists view racism as a toxic in the metaphorical sense, but I must add to that a much stronger attention to the literal, structural, and material aspects of racism's toxicity. This allows us to redefine racism itself not only as a pernicious form of social inequality but as a structural poison that kills.[34] Racism is a poison that fills the air, both figuratively and literally; it also fills the water and land literally, and it invades our bodies like the chemicals that TNCs manufacture and pump into our atmosphere every day. Toxic chemicals are the embodiment of racism (and gender and class violence) because they are intended to produce benefits for some while doing harm to others. But in the end, everyone is hurt to one degree or another. Environmental racism and the concept of racism as a poison converge to provide us with a language for redefining racism and deepening our understanding of this problem.

The poison or toxic metaphor for racism parallels Beck's risk society model in many ways. "Toxic," as Robert Bullard is fond of saying, "means poison."[35] Box 2.1 reveals a number of ways in which toxic chemicals and racism operate and cooperate.

We can conclude, then, that racism is a toxic behavior and ideology that has a negative impact on us all, but especially those who are its direct targets. Thus, we can deepen our understanding of environmental racism and rethink theories of racism through this lens.

It is also the case that to a large extent, class inequalities produce their own toxicities and reinforce social (including racial) hierarchies. Following Beck, Foster, Marx, and Schnaiberg, the domination of the environment is actually reflective of the domination of human beings, most specifically through class exploitation.[36] If environmental exploitation—which creates an inordinate volume of toxic chemicals each day—is also a system of class domination, then class domination is a toxic practice at its core. Class inequalities, particularly when they are deep and wide, produce pain and suffering among large populations of poor and working people (usually the majority). They can last for generations, creating the effect of permanent

Box 2.1
Racism as Toxicity: Racism and Chemical Toxins Compared

- Both can maim, debilitate, and kill their victims.
- Both are used to shape the earth and its landscapes literally and figuratively. Racist language and symbols have given us "the Dark Continent" of Africa and the "Orient," for example, and toxic waste has done untold physical harm to the landscapes of Africa, the Middle East, and Asia.
- Both are used to shape bodies literally and figuratively. For example, racist language and imagery in popular and scientific media have cast the bodies of people of color as inferior, prone to crime, oversexed, and unclean, while toxic poisons have physically deformed and debilitated countless persons of color in communities inundated with pollution.
- Both are simultaneously invisible and blatant. In the case of toxics, we can observe barrels of waste, garbage dumps, incinerators, smokestacks, oil slicks, and mine cavities, but the chemicals associated with these practices are often colorless, odorless, and undetectable by the human senses.
- Both migrate through space, over political and national borders, across social borders, through class, gender, and culture. For example, the United States–led war in VietNam saw the commingling of racial ideologies and the export of hazardous chemicals such as Napalm, Agent Orange, and dioxin, which migrated across national borders to do violence against nature and human bodies (see also Smith 2005).
- Both migrate through time, traveling across generations, with multiple generational impacts, and intensifying (bioaccumulating) throughout networks of interdependence. Toxic waste and hazardous chemicals can sit for decades in barrels, ponds, vacant lots, and other places and may take years to have a measurable impact on human populations. The mercury used to mine gold during the California Gold Rush continues to bioaccumulate throughout the food chain from fish to humans well over a century and a half after that event (Pellow and Park 2002). Racism leaves legacies and residues that can fester within individuals, families, communities, populations, nations, and regions over years, as evidenced in the continued struggle over the meaning of the Confederate flag in the United States and continued conflicts over the meaning of genocide in places like Turkey, Armenia, Iraq, the United States, Australia, Rwanda, and Kosovo, decades and even centuries after the fact. These forces shape consciousness, political alliances, national borders, and life chances, as well as the landscapes on which people live and depend.
- Both constitute what race theorists call social waste (Feagin, Vera, and Batur 2001) in that they reflect an unnecessary and destructive expenditure of societal energy and wealth.
- Both provide privatized benefits to certain people, institutions, and organizations, while imposing on and publicizing the costs to others. For example, states, corporations, and ethnically dominant populations share in

Box 2.1
(continued)

the immediate benefits of racism and environmental racism (although they also pay in the long run), while people of color suffer the immediate negative consequences.

• Both operate like a boomerang and circle back to haunt all members of society. Beck maintains that toxics in a "risk society" have a boomerang effect in that they loop around and harm those who produced them. This is in fact a fundamental principle of science and physics, or what Barry Commoner (1971) reminds us is the principle of "what goes up must come down." Many scholars and cultural critics find that racism also has its own boomerang effect—creating economic, psychological, and social/cultural discord and costs for whites (see Feagin, Vera, and Batur. 2001; Roediger 1999; Wise 2001, 2005).

• Both are powerful tools for organizing resistance movements and bringing people together across social and spatial boundaries. For example, the environmental justice movement consciously links racism, class inequalities, and industrial toxics for political mobilization purposes (see Gottlieb 1993, Pellow 2002).

poverty for certain groups;[37] despite their destructive social impacts, they can be quite difficult to detect;[38] they reinforce systems of ecological destruction; and they create resentments and tensions among the poor and working classes that periodically erupt in rebellion and—through environmental injustices and ecological boomerang effects—constantly produce toxic feedback loops that harm both rich and poor. Class inequalities can also provide a framework around which people within and across societies can organize to improve their lives and the environment they live in.[39] Poor and working-class people the world over—the global South majority—have a real stake in organizing against global North communities. But class politics is incomplete without an analysis of racial politics, and vice versa, and they must be joined with an environmental justice framework. But first a note on race theory.

Limitations in Research on Race

Extending the observations already offered, I offer two critiques of recent research on racism. The first concerns the political economy. There is ample discussion of how nation-states and governments engage in racial projects, producing discourses, ideologies, and practices that lead to and support inequalities between people of color and white Europeans and European Americans.[40] The state is also the critical target for reform efforts and social movements seeking racial justice. This is logical. However, what is missing from this state-centric approach is that European (and later American) capitalism, conquest, and slavery were largely made possible by the role and power of private corporations that financed and profited from all of these activities. The role of corporations is critical in understanding how populations and bodies were racialized centuries ago, as well as today.[41] This is important because unlike the state, private corporations are just that—private—and therefore not beholden to the vote or the will of the public as democratic governments are.[42] Corporations develop their own distinct cultures and practices with respect to how to engage racialized populations, whether they are customers, employees, or states. These cultures often conflict with official state policy concerning acceptable racial practices. More important, in the case of transnational corporations, these organizations are, by definition, not subject to the will of a single state because they operate in multiple geographic spaces and legal frameworks. In such cases, the role of the state is necessarily diminished, and the corporation may take on a global culture or multiple cultures in relation to racialized populations.

The second critique is that there is a good deal of discussion about the structural impacts and effects of racism on people of color in the United States and globally, but the literature on racism has yet to seriously consider the ways in which environmental and natural resource destruction is embedded in institutional racism. Consider Barlow's emphasis on materiality: "The process of racialization is fundamentally a *material* relationship. Racial privilege refers to people's capacity to make unequal claims to scarce social resources—for example, freedom, citizenship, jobs, political power, housing, education, and prestige—because of their racial designation. The always-present flip side of racial privilege is racial subjugation (or oppression): the denial of equal access to scarce social resources on the basis of racial designation."[43] Natural resources are absent from this framework. From an environmental sociological perspective, ecological and natural resources could easily and logically be considered an example of "scarce resources," and given the extent of environmental inequality within and across nations, those resources are certainly racialized. Natural resources are used and abused to support racial hegemony and domination and have been at the core of this process for a half-millennium. For example, when African American employees at a Fortune 500 company like Texaco file a racial discrimination suit against the firm, we tend to view this incident in familiar terms: civil rights, institutional racism, employment discrimination. All of these are useful, but unless we go beneath these practices and understand the material and structural sources of Texaco's power—beyond its revenue and profit margins—we learn only part of the story. From an environmental standpoint, Texaco's discrimination against its African American employees is supported by a foundation of global natural resource extraction and environmental degradation associated with the fossil fuels it extracts, refines, and sells, harming indigenous peoples' health, land, and watersheds in places like Ecuador, and contributing to air pollution and climate change more generally through its refining and transportation operations. Texaco eventually paid out more than $100 million in a legal settlement over charges that it had discriminated against its African American employees.[44] It has only recently been forced to consider doing the same for environmental and human rights abuses alleged in other nations.[45]

Without attention to the material and natural resource base of this company's wealth, we achieve a limited understanding of the foundation of institutional racism. The exploitation of human beings and the natural environment are linked, and this is one such example. Theoretically, we can then rethink our models and ideas about what racism itself is. Typically we think of racism as "a system of oppression of . . . people of color by white Europeans and white Americans."[46] But the material sources and the structural foundation of such power and manipulation are contained in the natural world: timber, oil, land and soil, seeds, crops, minerals, metals, and water. The conquest of Native peoples in the Americas was primarily a racial project characterized by the domination of people and natural resources such as gold, sugar, and spices. Slavery was a system built on the conquest, characterized by African peoples working under coerced conditions on land stolen from aboriginal peoples. The industrial revolutions in Europe and the United States were also fueled by human exploitation (slave and free wage labor) to produce agricultural products and factory goods for consumption and export around the globe. The twentieth- and twenty-first-century economies are not fundamentally different in that the manufacturing and service sectors are marked by extreme wage and salary differentials along race, class, gender, and immigration status lines and rely on a greater volume of natural resources than ever before.[47]

The late Ogoni poet and environmental justice activist Ken Saro-Wiwa referred to the destructive impacts of multinational oil corporations in his homeland in Nigeria as *omnicide,* a term he coined to capture the full scope of the combination of ecocide and genocide. He once wrote: "Men, women and children die unnoticed, flora and fauna are threatened, the air is poisoned, waters are polluted, and finally, the land itself dies."[48] *Omnicide* is a powerful way to encapsulate what environmental racism is, and it is also a dire warning to all of us because we all inhabit a global risk society in which we are implicated or in some way burdened by the destructive power of late modern capitalism. What this suggests is that any comprehensive analysis of class and racial politics must be integrated with an environmental justice framework.

Toward an Environmental Justice Analysis of Class and Racial Inequalities

This section offers a synthesis of theories of class struggle and environmental conflict and connects these ideas to theories of racial inequality. The goal here is to produce a more coherent environmental justice analysis by demonstrating the threads that unite these ideas in a way that avoids pitting race against class. Schnaiberg et al.'s treadmill of production is a good starting point (see chapter 1). Within the treadmill, the goal of corporate profit making exists in fundamental tension with the goals of environmental protection and social and economic welfare for the working classes. That is, capitalism is incompatible with ecological sustainability and economic justice. Thus, if environmental harm is linked to the exploitation of the working classes, then environmental sustainability might be tied implicitly to working-class empowerment. Under such a system, the working classes have an interest in environmental sustainability.

Paralleling the approach of the treadmill of production, Ulrich Beck's risk society thesis argues that corporations hold a great deal of power in societies. He emphasizes the role of corporations in the control over economic development and scientific institutions, which poses difficulties for producing equitable economies and progressive scientific practices. Beck perhaps places too much faith in the promise and power of scientific experts, whom he argues are critical to deciphering and translating the problems of a risk society. This is problematic, of course, because it maintains the marginality of working-class populations in this process. And since, according to Beck, risks "adhere to the class pattern,"[49] which means that working-class and poor communities are those facing the greatest exposure to environmental harm, these populations would likely need to be much more involved. Accordingly, Schnaiberg et al. argue that progressive environmental, social, and economic change will likely only result from massive disruptive action from below. Thus, Beck's framework is productive as far as his understanding of corporate power and environmental inequality, and Schnaiberg et al.'s framework complements Beck in that it offers a deeper understanding of the origins of the problem and a potential path forward in class terms.

Class (like race) reflects not only social position but relationships and conflicts among groups vying for resources.[50] The treadmill of production and risk society theses both make this process clear. The global political economy (and most domestic systems) supports a framework wherein class privilege affords one access to spaces where wealth flows and pollution is minimal, while vulnerable majorities contend with toxics and poverty. State and corporate policies function to produce and reproduce these inequalities that Beck writes will eventually haunt all members of society. According to Schnaiberg, such justice occurs only when social movements present disruptive challenges to these systems. Class inequalities are produced and challenged through class struggles.

In socioecological terms, Karl Marx argued that class inequalities in capitalist systems reveal the fact of the domination of both nature and humanity, and the two are coupled through regimes of private property that commodify all living things.[51] In other words, human beings and nature lose their inherent worth and are defined mainly in market value terms. For indigenous persons, people of color, and women, such a system is additionally burdensome because their "market value" (the wages and salaries they can command) is further degraded as a result of ideologies that define them as socially and culturally worth less than others. Likewise, the spaces—land, neighborhoods, workplaces—where these populations live, work, and play are degraded as worth less than other spaces, providing a political-economic logic for environmental inequality. Thus, class inequalities intersect with and reinforce racial and gender inequalities, creating complementary ecological impacts as well.

Class inequalities within and among nations also produce their own toxicities and reinforce racial hierarchies at multiple scales. In fact, all environmental injustices are the result of a combination of racial and class (among other) inequalities. As Stuart Hall notes, for people of color, class is lived through race,[52] so that even when China, India, Japan, Brazil, or other non-European nations achieve something close to developed nation status, they are still racialized differently, and this has significant consequences for how these states and their citizens are viewed and treated in relation to more dominant nations. Specifically, when and if these nations acquire considerable military or economic power, they are often also viewed by northern nations as threats to the cultural, racial/ethnic, or religious fabric of the global order, so there is an additional burden of being racialized differently (and as inferior) that comes with being an ascendant state populated by people of color, regardless of class/economic or military position. For example, Japan is a member of the Group of Eight nations, yet despite its global economic influence, numerous hate crimes are committed against Japanese citizens and Japanese Americans every year on U.S. soil.[53] Citizens and descendants of other Asian nations face a similar plight: the number of hate crimes facing South Asians and South Asian Americans is much higher than that for most other groups in the United States.[54] Thus, while the Asian Tiger economies have considerable influence on global financial systems, to a great extent the East is and will always be foreign and alien from the West's point of view and must never be allowed to enjoy dominance. After all, European economic, cultural, political, and military dominance has been a global phenomenon for a half-millennium, while Asian economic power has largely been confined to that region of the globe. Viewed through this lens, the rise of certain Asian economies is perhaps a global version of what Yen Le Espiritu has called *differential inclusion:* "The process whereby a group of people is deemed integral to the nation's economy, culture, identity, and power—but integral only or precisely because of their designated subordinate standing."[55]

My point here is to recognize the importance of class as a driving factor in global environmental politics—particularly when the perpetrators of environmental inequalities are global South states and corporations—but also to underscore that these class dynamics are never separate from global racial hierarchies and ideologies. Following Guinier and Torres, my analysis moves from race to class without ever losing sight of race.[56] That is, I include both measures of inequality and power to demonstrate how they intersect, but my general framework places a greater emphasis on the lens of racial inequality.

The environmental and social dimensions of the modernity project discussed above do not exist in separate domains. In fact, they are not even simply parallel phenomena, but rather they have been consciously intertwined by those who benefit from these practices. Modernity is a project that is deeply racist, classist, and patriarchal and quite destructive ecologically, and this is by design. What is being done to address the socially and ecologically harmful practices associated with modernity? Transnational social movement networks are creating spaces where formidable resistance efforts have taken hold.

Transnational Social Movements

Local and national social movements have been a primary scholarly emphasis since sociologists began studying protest behavior in the nineteenth century. Given the great proliferation of transnational movements, recent scholarship has paid more attention to that phenomenon than ever before. The study of social movement organizing at the transnational scale began in the 1970s and grew into a major field of research by the late 1980s and 1990s.

Social movements result when networks of actors relatively excluded from routine decision-making processes engage in collective attempts to change some aspect of the social structure or political system.[57] Critical components of all successful social movements include organizing and mobilizing resources, framing grievances and goals, and engaging the political opportunity structures that constrain or enable social change. In the following sections, I focus mainly on organizing/mobilizing and political opportunity structures. I argue that transnational social movements confront a global political economic opportunity structure shaped by racial, class, and national inequalities and that resistance against institutions creating transnational social and environmental inequalities produces new spaces for the articulation of global citizenship. This is what some scholars call a "transnational public sphere"[58]—a space in which activist networks engage each other and their targets across national borders and articulate a vision of global justice.

Mobilization and Organization

I draw mainly on sociological theories of social movements because they are the conceptual tools most useful for studying political mobilization by activists, their organizations, and networks. Theories of agency and resistance in disciplines like postcolonial studies tend to embrace a much broader, complex framework for understanding movements than does sociology. Postcolonial studies view power and resistance in multiple and often contradictory forms rather than focusing on resistance mainly through more dramatic and publicly confrontational venues.[59] Social movement theorists tend to ignore the uneven, complicated, and messy nature of power relations, while postcolonial and cultural studies scholars embrace it.[60] While positioning this study principally from a social movement theoretical standpoint, I am influenced by both of these schools of thought in order to capture the dynamism of power relations among movement groups and between activist networks and their opponents.

Movements are rarely successful if they do not possess the ability to engage in disruptive action.[61] Sidney Tarrow calls this "contentious politics"—the act of "confrontation with elites, authorities, and opponents."[62] I argue that when social movements challenge official discourses, definitions of the situation, and monopolies on information production, they are also engaging in disruptive action. Disruptive action need not always involve physical confrontation and mass mobilization against authorities, although that kind of action is often necessary. But building social movements requires that activists move those contentious politics beyond the episodic and explosive engagements that characterize so much of "politics by other means," and coordinate and sustain this kind of work over time. Tarrow contends that these kinds of social movement behaviors are synonymous with the rise of the modern nation-state.[63]

Nation-states are at the core of transnational social movement organization (TSMO) activities. TSMOs are defined as "a subset of social movement organizations operating in more than two states."[64] These organizations coordinate action to address problems that affect people in two or more nations and are reflective of the increasing global interdependence of societies. TSMOs specialize in "minding other people's business"[65] and are largely based in the northern hemisphere for reasons of access to global decision makers in London, New York, and Washington, D.C., but also because telecommunications and transportation in these nations are often more supportive of rapid and intense utilization by activists who may need to communicate or mobilize on short notice. And although the leadership of most TSMOs is based in the global North, there is a rise in TSMOs in the South, reflecting the growing understanding that much of the primary leadership in global justice movement efforts must be based there.[66] Complicating the North-South binary in TSMO organizing, a number of environmental justice and human rights activists from the global South live and work for TSMOs in the North.[67]

TSMOs have been proliferating since the end of World War II,[68] the moment when economic globalization intensified through the rise in transnational corporations (many of which produced and used chemical compounds that pervade the environment)[69] and when the historic break in global race relations occurred.[70] These three developments are interrelated. As TNCs competed and collaborated with nation-states through neocolonial arrangements in the global South (giving rise to the new global economy), TSMOs rose up to combat corporate and state violations of human rights within and across national borders, emboldened by the credibility of the United Nations when it announced the Universal Declaration on Human Rights in 1948. TSMOs are now widely acknowledged as formidable players in international politics as they create new global norms and practices among states, international bodies, and corporations and transform old ones.[71] Moreover, these nonstate actors are viewed as sources of resistance to neoliberal globalization.

The mobilization of human bodies in protest is something that occurs primarily and most effectively at the local level and in particular places. And although many local and national social movement organizations (SMOs) and even TSMOs can accomplish this in multiple places simultaneously,[72] in the case of TSMOs—which are, by definition, extralocal and not actually rooted in one particular place—the greatest resources at their disposal are information and the ability to produce, communicate, share, and strategically use it as leverage. A growing and dense web of cyber, telephone, and fax networks has positioned TSMOs as challengers to the information monopolies that many nation-states enjoyed a short time ago.

TSMOs are rarely successful if we narrowly define success as a major change in a specific policy within a nation-state.[73] But they are increasingly relevant in international policy debates, as they seek not only to make policy changes but also to change the terms of important debates. TSMOs are often a critical source of information for governments seeking to learn about a problem, and their presence raises the costs of failing to act on certain issues, thus increasing government accountability. In a global society where a nation-state's reputation on a host of matters can be tarnished in international political and media venues, TSMOs can have real influence. They use the transnational public sphere as a venue through which to hold state practices up to international community standards, scrutiny, and judgment. Human rights activists have called this tactic the "mobilization of shame."[74] Because few governments, even democratic states, allow broad public participation in their foreign policy processes, most international decision making occurs with little public accountability. Thus, TSMOs serve as a vital resource for domestic and transnational social justice.

North-South NGO Politics and Southern Environmental Justice Organizing

As I noted in chapter 1, a significant factor driving the globalization of hazardous waste dumping was the passage of more stringent environmental legislation in northern nations—an indicator of the impressive power of northern environmental movements. This unfortunate outcome of growing grassroots northern environmentalism revealed the need for environmentalists to think and act critically, locally, and globally.[75] Thus, many of the transnational environmental movement groups and networks based in the global North developed not only as a means to combat growing global environmental problems but also, in part, as an ethical obligation to take responsibility for hazards generated in and exported from the global North. They recognized that ecological modernization in the North was often necessarily coupled with environmental racism and inequality in the South: "Many of the leading environmental organizations realized that environmental victories in northern industrial countries could easily be negated by an expansion of polluting industries in the Third World. Environmental problems such as deforestation, acid rain and climate change were global problems requiring global solutions. This emerging global perspective was reflected in the fact that it was precisely those environmental groups that concentrated on international issues that grew most successfully throughout the 1980s."[76]

This was easier said than done, however. Scholars and activists have noted for some time that a recurring concern among activists and residents in the global South is that TSMOs (and, by extension, transnational civil society itself) overwhelmingly represent the concerns of northerners, who are endowed with greater resources and often have distinct interests and histories.[77] This tension mirrors the conflicts scholars have observed between white, middle-class environmentalists and environmental justice activists from working-class backgrounds and communities of color in the United States.[78] In order for North-South alliances to be successful over the long term, they have to be more than just "fragile fax-and-cyberspace-skeletons" that lack any long-term direction.[79] They must involve respect and trust. This does not come easy.

Movement scholars argue that TSMOs facilitate the development of a global civil society by providing local and national SMOs around the world access to leverage and power in relation to their subnational and national governments that they might not otherwise have.[80] TSMOs offer a civil society presence at international treaty fora and policymaking venues and therefore might serve as a democratizing influence on global civil society. While that may be true to a great extent, some scholars may overstate the affinity and solidarity between groups in the global North and the South. The hopes for global civil society must be tempered by the experiences of local and indigenous groups battling racism from within environmental groups and other TSMOs.[81] Activists from the South often see their own goals and vision of social change take a back seat to the directives and agendas of northern groups, and this produces conflict and distrust. High on the list of priorities should be the development and maintenance of relations of trust, respect, and power sharing as critical to holding such transnational activist networks together.[82] In 2003 a global network of southern activists released a public statement on the tensions between northern nongovernmental organizations (NGOs) and southern groups. Calling themselves People against Foreign NGO Neocolonialism, they presented numerous critiques of northern environmental TSMO dominance over southern environmental justice issues.[83] The following are selected quotes from that report:

"I do not need white NGOs to speak for me."—James Shikwati, Africa

"Some call it 'environmental colonialism,' others call it plain racism and privilege. The underlying problem is often quite basic, revolving around historic views of who should control the land, perceptions of Native peoples, and ideas about how now-endangered ecosystems should be managed."—Winona LaDuke, Native American activist

"Under the banner of saving the African environment, Africans in the last few decades have been subject to a new form of 'environmental *colonialism*.'"—Dr. Robert H. Nelson

"Citizens who have become CI [Conservation International] country directors are right in the middle of continuing this *colonial* system because they refuse to rock the boat enough so the necessary change will occur. Don't ever trust these foreign NGOs, enough of their secret memos have leaked by now to show they always have a hidden agenda."—Name withheld on request

As these activists have noted, they draw no hard distinction between state and corporate colonialism and environmental colonialism because, for them, it is yet another form of exploitation by northern forces. These protest voices from the South echo U.S. environmental justice activist Dana Alston's famous words, "We speak for ourselves," which became a rallying cry for the U.S. environmental justice movement after numerous tensions erupted between activists of color and white environmentalists over a host of cultural conflicts and disagreements concerning the distribution of movement resources.[84]

Scholars argue that transnational social movements are critical for lending legitimacy, visibility, and power to local community protest efforts in nations and regions where authorities might otherwise ignore such concerns entirely. This is often the case. But as the above quotations reveal, northern TSMOs must be willing to work on repairing the existing damage in North-South SMO-NGO relations if global civil society is to truly challenge the global status quo.

From Political Process to Political Economic Opportunity Structures

Two of the major dimensions of social movement theory focus on mobilization and political opportunities. Having considered questions of mobilization in the previous section, I now focus on political opportunity structures.

The political opportunity structure (POS) model of social movement conflict emerged with Peter Eisinger's research on urban rebellions in the United States.[85] Eisinger concluded that these conflicts occurred most often in cities with a mixture of open and closed governmental structures. Since that time, scholars have developed Eisinger's earlier framework into the political process model[86] and the international political opportunity structure model,[87] which have been useful for explaining the inception, development, and suppression of social movements. By explaining collective action as the result of factors internal and external to social movements, the political process model represented a step beyond traditional resource mobilization theory, which focused primarily on internal movement factors. Movement scholars had to consider the broader political context.

The political opportunity structure consists mainly of the following dimensions:

- The relative openness or closure of the institutionalized political system.
- The stability of elite networks that typically underlie a political system. That is, if elites are divided, this may be an opportunity for social movements to exploit.
- The presence of elite allies who sympathize with or support social movements.
- The state's capacity and propensity for repression.[88]

Distilling these factors, Keck and Sikkink define the political opportunity structure as "differential access by citizens to political institutions like legislatures, bureaucracies, and courts."[89] In other words, movements must engage the state in order to bring about lasting change.

This model is deeply wedded to a state-centric perspective.[90] As Boudreau writes, "It is generally assumed that all movements seek access to state institutions and decision-making power."[91] The state is the principal movement target or primary vehicle of reform. Although many movements do exhibit this state-centric approach, others increasingly view the nation-state as just one of many targets in their campaigns for social change. Thus, the principal foci and site of reform are no longer centered within states. The state or political elites are only one segment of the political sphere and are sometimes sidestepped by movements when they target other powerful institutions, such as corporations.

Since the Bretton Woods Agreement of 1944, the emergence and convergence of numerous international banks and financial institutions (IFIs), transnational corporations (TNCs), and free trade agreements have produced profound transformations in the global political and economic terrain. Bretton Woods was a meeting of U.S. and European allies who created the World Bank and International Monetary Fund (IMF) to stabilize currencies and facilitate conditions for capital investment and export in global South nations. This laid the groundwork for free trade agreements that would facilitate the needs of banks and corporations for decades to come. Free trade agreements are binding arrangements that establish supranational limitations on any nation's legal and practical ability to subordinate commercial activity to the nation's goals. In that sense, they are the embodiment of the profit-before-people orientation that global justice activists decry. IFIs like the IMF and World Bank provide loans and funds for development projects to nations in the South, but generally only on the condition that they restructure their economies, laws, and priorities toward the single goal of repaying the loans. Dubbed structural adjustment programs (SAPs), these changes included shifts from staple crops to exotics for export to earn hard currency; cuts in education, health, and other social spending; and wide-scale divestiture of public assets. The result has been widespread hunger, environmental degradation, poverty, inequality, and debt.[92] Nations throughout Africa, Asia, Latin America, and Central and Eastern Europe have been saddled with enormous debt from IFIs and have hung their hopes for recovery on opening up their natural resources and consumer markets to TNCs. SAPs and free trade agreements have proved to be a disastrous combination in the South. Almost all global South nations that have adopted trade liberalization schemes have experienced significant wage inequalities.[93] Many citizens worldwide are now just as likely to blame an international bank, a corporation, or a free trade agreement for these problems as they are their own governments.[94]

Given these significant changes in the past half-century, I propose an extension of the political process model: the political economic process perspective.[95] This model acknowledges the intimate associations between formal political institutions (e.g., states and legislative bodies) and economic institutions (e.g., large corporations and banks) and their engagements with social movements. The political economic opportunity structure stresses the extensive influence of capital over nation-state policymaking, regulation, and politics and views corporations as equally likely to be the targets of social movement campaigns.[96] This model also underscores that divisions among corporate elites are just as important as divisions among political elites (the latter being a primary focus of the classic POS model).[97]

The political economic opportunity structure model offers a connection to the study of TSMOs precisely because these groups have often targeted private sector actors, particularly TNCs, from their inception. Unlike students of domestic and national social movements, many scholars studying TSMOs have incorporated nonstate organizations like private sector institutions into their models as social movement targets.[98] TSMO targets can include multiple governments, IFIs, and private actors like TNCs. Yet what is still missing from the literature is an integration of private corporations into the political opportunity structure model itself. Consider Smith, Pagnucco, and Chatfield's conclusion that "political opportunity for TSMOs is structured by the national, intergovernmental, and transgovernmental arenas in which they operate."[99] In other words, the political opportunity structure in which movements maneuver is still bounded by states. Recently a few scholars have begun to theorize the role of industrial actors in the political process, but these studies have only scratched the surface with regard to the power of capital, its relationship to nation-states, and the implications for social movements seeking environmental justice and human rights.[100]

As powerful as nation-states and TNCs may be, it is always critical to remember that social movements not only engage political economic opportunities; they can also create and open them. One of the most effective ways to create access to domestic political systems is through international pressure. When local activists and advocates abroad make an issue visible to the rest of the world, it can create a "boomerang effect that curves around" local nation-state indifference and oppression to place pressure for policy changes on these local governments (and corporations).[101] Thus, the international political opportunity structure frequently consists of a combination of domestically closed structures with more accessible structures elsewhere in places where TSMOs may be based. When we integrate a political economic opportunity structure perspective, the interactions among movements, states, and corporations create additional openings, opportunities, and points of access for activists to exploit, ultimately redefining transnational politics and the transnational public sphere.

The reality is that social movements remain deeply concerned about state power, but they are frequently just as concerned (if not more concerned) about corporate power. For evidence, one need look no further than the rise of the corporate campaign phenomenon, whereby activist networks target one or more large corporations to change their policies on a series of issues. The campaigns against the Citibank and Home Depot corporations targeted them for their support of destructive timber harvesting practices; the Newmont Mining Corporation faces a global campaign that places the spotlight on its socially and ecologically harmful mining operations on at least three continents; Shell Oil confronts a coalition of U.S., European, Asian, and African activists pushing for changes in its policies concerning the environmental and economic impacts of its petroleum extraction and refinery operations; and in the past several years, Dell, Hewlett-Packard, Apple, and Compaq computer corporations have all faced sophisticated activist campaigns that have forced each company to compete with the others over the design of more sustainable computer production and recycling systems. Whether any of these goals will lead to actual environmental and social justice is another question. The point is that tactics have shifted, and movement scholars have taken little notice.

While many scholars underappreciate the power of corporations, others have considered the adaptive role the state plays in the face of capital.[102] In fact, Bakan argues that the state's powers have not actually receded in the face of capital's dominance; they have only shifted and even expanded away from protection of public goods to the protection of corporations and private goods.[103] This shift corresponds to state practices that restrict public services from new immigrants and people of color who are viewed as having less than a legitimate claim to them.[104] Such shifts hurt all working-class and poor people as well.

A political economic opportunity structure approach by social movements that targets states and corporations responsible for social exploitation and pollution is critical for achieving local and global environmental justice. But the focus on corporations is also problematic because, as Bakan argues, it may lead to the privatization of protest or, as O'Rourke contends, to the privatization of regulation.[105] Either way, that strategy ultimately runs the risk of ignoring the crucial role of the state, which is ultimately necessary for ensuring democratic processes and the protection of public welfare. Moreover, most movement organizations running corporate campaigns tend to shy away from presenting a fundamental challenge to the market economy's principle of growth at all costs. Instead, these campaigns generally seek a reduction in the negative impacts of production rather than confronting the deeply antiecological nature of the market economy itself.[106]

Why does this matter? The importance of understanding the tensions between state and corporate power in national and global politics is critical because states are a form of governance (at least in democracies) that are supposed to be accountable to the citizenry. Corporations are not. The only constituencies to which they are accountable are the firm's shareholders. In fact, they can be in total compliance with the law when they make decisions that penalize workers, communities, and consumers, as long as their shareholders are satisfied. So if institutions of this ilk are making decisions about the public good and the general welfare of a population and its environment, we have entered into an era marked by a troubling reality. This is particularly acute in areas where poor, working-class, immigrant, and people-of-color populations are concentrated in economically deprived and politically less powerful enclaves that are frequently burdened with the toxic cast-offs of production and urban consumption patterns. TNCs exist within and beyond national boundaries and legal structures (and can influence the adoption of new laws and push states to change old ones). In such cases, the role of the state is undermined or exploited for private gain. Corporate power today is as important as the power of states as sites of environmental policymaking and the production of social inequality, and their contestation. Thus, social change in this context requires a different set of tools and strategies.

Ultimately the debate over whether states or capital have more or less power is counterproductive, because states have adopted a corporate form.[107] That is, it is conceptually a mistake to even separate the state from capital because: (1) so many powerful political actors are also economic actors who own stock and interest in corporations and have personal stakes in corporate profits;[108] (2) states make economic decisions every day, and corporations make political decisions routinely; and, most important, (3) states increasingly mimic corporate behavior and embrace ideologies of privatization.[109] For years, federal, state, and local governments in the United States have reorganized themselves internally in how they approach the daily task of governance and service delivery to conform to a business model. This comes in many forms:

- Corporate-style performance evaluation procedures for employee promotions
- Outsourcing and privatizing core public functions (education and health care, for example), privatizing the commons (public space, land, air, and water), and downsizing governments to achieve greater accountability and efficiencies
- City mayors describing themselves as "CEOs," as San Diego mayor Jerry Sanders does
- Elected officials using language like "the contract with America" that reflects a business sensibility much more than it does any sort of social contract
- Transnational corporate form actions, like the U.S. Department of Agriculture's deputizing Mexican nationals to perform its produce certification functions inside Mexico before these goods cross the border into the United States[110]

E. H. Isin argues that states have redefined the citizen as clients and market actors responsible for their own lives.[111] In other words, many states have taken on a corporate form and have blurred the line between government and industry entirely. It appears that TSMOs are well aware of this change, as they articulate a political economic opportunity structure framework that informs collective action. As Greenpeace recently wrote to supporters and members, "We resent the *companies and politicians* who put profit before the health of our planet."[112]

Sustaining Disruptions, Securing New Openings

There are numerous other methods TSMOs use to create and open political economic opportunity structures, which raises the question as to how much power TSMOs possess. For example, when TSMO activists influence the adoption of international conventions, these are political opportunities that can open additional windows of access for local movement action. International norms constitute critical resources from which TSMOs draw their ideas, while at other times, "transnational networks, coalitions, and movements may attempt to transform their collective beliefs into international norms."[113] Thus, it works both ways. For example, the United Nations General Assembly approved the Universal Declaration of Human Rights in December 1948, and the international community built on this declaration and reinforced it with two legally binding international covenants on human rights that entered into force in 1976.[114] Human rights groups were involved throughout this process. The Stockholm Convention on Persistent Organic Pollutants was developed by nation-states and environmental organizations and then used as a tool of persuasion and embarrassment against nations that had not yet ratified it (as well as against those that had but were in clear violation of it). When it finally went into force in 2004, activists who had worked so hard to get the convention to that stage breathed a short sigh of relief, then immediately used it to push for environmentally sustainable practices in nonmember nations. Thus, activists create international norms and then work to apply them around the world.[115] One of the principal aims of transnational movement activity, then, is to develop, implement, and track international norms. The Basel Action Network, Global Anti-Incinerator Alliance, Greenpeace International, Health Care Without Harm, International Campaign for Responsible Technology, International POPs Elimination Network, International Rivers Network, Rainforest Action Network, Silicon Valley Toxics Coalition, and many others have all emphasized this approach.

At times, environmental and human rights activists have had remarkable success at injecting the language of environmental justice into international conventions. In 1994, the UN's Draft Declaration of Principles on Human Rights announced the universal human right to a "secure, healthy, ecologically sound environment."[116] And in upholding the right of indigenous peoples in Ecuador to sue the Texaco oil corporation for environmental damages to their land, a U.S. judge cited the 1992 Rio Declaration, which—as a result of the work of environmental activists from all over the world at the Earth Summit—declared access to a clean and healthy environment to be a fundamental and inalienable human right. Transnational environmental movement efforts have achieved their greatest success when they have made the connection between protecting critical ecosystems and protecting the people who inhabit such spaces.[117] This is what the movement for environmental justice has stood for since its inception: the inextricable relationship between the degradation of people and their ecosystems. When movements can articulate these links and integrate them into international norms and state and corporate policies and practices, this constitutes a remarkable achievement, because it involves both discursive and structural "disruptions" in the otherwise normal flow of power.[118]

Race and Transnational Social Movements

In the literature on transnational social movements, there are two shortcomings of relevance to this study. The first is the lack of theorizing about the ways in which private corporations routinely exercise hegemony over domestic and transnational political processes. We have addressed this issue above. The second shortcoming is that race and racism are all but totally absent as motivating factors in the development of transnational collective action and in the conceptualization of globalization, the transnational political process, and global civil society. A review of nearly all the major works in the field of transnational social movement studies reveals how widespread this trend is. Given how deeply racialized many domestic and transnational social movement struggles are, this oversight is surprising.

For example, Keck and Sikkink and their collaborators present case studies of transnational movement campaigns targeting female foot binding in

China, female circumcision in Kenya, environmental activism in the global South, and antislavery campaigns in the United States and the United Kingdom. It is surprising that considering how clearly racialized each of these issues and campaigns is, the analysis by social movement theorists neglects the significant factor of race. Slavery abolitionism is believed to have been perhaps the first major transnational human rights movement. This movement involved people of color and women speaking in public for the first time in Europe and pioneered the corporate campaign and boycott tactics, as well as produced the first mass-published activist newsletters.[119] This was a movement with roots in religious organizations that sustained a long-term campaign to oppose a system characterized by racial injustice and human rights abuses.[120] If indeed this was the first transnational human rights movement, I believe this suggests something unique about racial justice efforts and their appeal to people across national boundaries. Yet movement scholars have paid little attention to this aspect of that movement and others.

Transnational political economic opportunity structures themselves are deeply racialized (in addition to being imbued with class dynamics, of course). Modernization theory, for example, pervades both the academy and international politics and, in essence, is the position that "developing" and "preindustrial" nations are in need of assistance and guidance from "developed" and "industrialized" states. This viewpoint ignores the hundreds of years of underdevelopment of the South by the North, as labor and natural resources were exploited for the benefit of European colonizers in a process legitimated by overtly racist ideological perspectives that positioned whites above people of color.[121] Yet today many scholars of globalization seem unaware of this history. While analyzed in terms of global social inequality, class divides, and gender exploitation, global inequalities are rarely framed as racial inequalities.[122]

Consider the tensions among nations in Asia and Africa in the mid-twentieth century, as those parts of the world were becoming newly independent of colonialism. One of the primary reasons that the U.S. government gave concessions on its position regarding legalized segregation of African Americans was the pressure brought on by heads of state from some of these nations. Many of these leaders—and TSMOs and NGOs working in solidarity with them at the time—pushed for an integration of racial justice initiatives into international law. Language addressing racism as a global human rights concern has long been a part of international conventions.[123]

The remarkable work that TSMOs have done to develop human rights accords provides access for people in the global South to raise issues regarding their historical and contemporary marginalization. The development of global rules and conventions on racism and human rights reveals that these are critical sites for mobilization by nations and peoples who have undergone slavery, colonialism, and neocolonial racism.

International law and norms, combined with the rise of transnational corporate power, are two developments that have contradictory effects on states and publics. The development of international law allows citizens in any nation to potentially seek protections against abuses that their own government might not otherwise provide them. The rise of TNCs is a countervailing force that produces abuses against publics and also limits state powers. For publics and social movements, the first development is a blessing, while the second is a curse. For states, the first development is often viewed as cumbersome and a threat to sovereignty, while the power of TNCs is viewed as facilitative of state power (although it can threaten state sovereignty as well).

The question of state sovereignty has deep racial dimensions as well. For many northern transnational activists, although the erosion of state sovereignty in their hemisphere at the hands of corporations is seen as deleterious (because it compromises the rule of law and citizen-led democracy), the same dynamics in the southern hemisphere may often be viewed as an opening in the political economic opportunity structure because it is assumed that many such governments are unelected or are run by corrupt officials: "Northerners within networks usually see third world leaders' claims about sovereignty as the self-serving positions of authoritarian or, in any case, elite actors. They consider that a weaker sovereignty might actually improve the political clout of the most marginalized people in developing countries."[124] However, for citizens and activists in the South, this is a different matter altogether, as many view the doctrines of sovereignty and nonintervention as the main paths to national self-determination, even if it remains elusive.

This dynamic has racial implications because when northern activists take the position that the erosion of southern sovereignty is an opportunity, it assumes that many global South nations cannot govern themselves properly without external guidance and intervention,[125] and it also ignores undemocratic practices in the North and the historic and ongoing impact of northern state and corporate colonization on southern nations' political and economic structures (for example, the northern state and corporate support of corrupt regimes and overthrow of democratic ones).[126]

This "southern states as corrupt" perspective is reflected in much of the research on transnational movements and global civil society. As one study states, "Of course, there are deviations from the norms of world society such as human rights violations or the cronyism of politics in many developing countries."[127] Such statements are often made without reference to a specific nation, government, or instance in which corruption was documented or observed; rather, they are just sweeping references to "Third World governments" as beset by corruption. Northern governments are often just as resistant to change and as secretive and corrupt on a range of issues as any other state and may require as much scrutiny, persuasion, and transparency.[128] To assume otherwise and to focus on global South governments as inherently corrupt is deeply problematic—and inaccurate. Hence, political economic opportunity structures are very much racialized.

Conclusion

Studies of race and racism pay little attention to the role of environmental politics in shaping racial inequality and to the power of corporations in the production of national and global racial logics. The historic break in race relations began at the end of World War II when TNCs emerged as a new political and economic force around the globe—a force that maintained and transformed racial and class inequalities and environmental exploitation. This resulted in the growth and dominance of corporations over the planet and its peoples and demonstrates the toxicity of racism and class domination.

The scholarly literature on transnational social movements and global civil society contains two principal gaps of concern here. One is the absence of an integration of political economy into models of opportunity structures. The other is the silence of scholars on the role of race and racism in the development of TSMOs and the normal operation of the global economy and geopolitics. Racial dynamics, of course, explain only part of the picture, but they are significant and overlooked almost entirely by students of transnational movements. Both observations are of critical importance in the study of transnational corporate activities involving the export and dumping of hazardous wastes from northern nations to southern nations, the latter being populated primarily by people of color and poor persons. These points are also noteworthy because they shape the political economic environment in which local and global movements for environmental justice exist and operate.

The activists featured in this study target the seats of power in each particular struggle, whether nation-states, corporations, or both. Transnational activists are changing the terms of debate over environmental protection and human rights, placing new and progressive ideas on the agendas within and across nation-states and challenging the neoliberal model of economic globalization. Transnational social movements frequently force nation-states and corporations to respond to their citizenry and consumers, thus articulating a new form of global citizenship that combats both traditional narrow models of national citizenship and more recent notions of corporate citizenship.[129] This means that corporations and states are potentially accountable to citizens in other nations and those citizens' governments in ways rarely witnessed before. They are also accountable to third force groups—social movements that operate to mobilize shame in an era when opinions of the international community matter more than ever. In fact, TSMOs constitute a core part of that international community. The development and strength of transnational social movements suggest the possibility that in the future, the transnational public sphere will not be restricted to seasoned activists, but rather will be much more open to citizens everywhere, transforming the very idea of citizenship in a global society.

3

Transnational Movement Networks for Environmental Justice

In this chapter, I consider key actors in some of the major transnational environmental justice networks operating around the globe today. I present some of the core concerns, controversies, and strategies these activists, organizations, and their networks confront as they build a movement to contend with the forces of global inequality and environmental harm.[1]

Transnational Social Movement Organizations and Networks

We begin with an introduction to some of the major networks that form a significant constellation within the broader global environmental justice movement.[2]

Basel Action Network (BAN)

BAN describes itself in this way:

> An international network of activists seeking to put an end to economically motivated toxic waste export and dumping—particularly hazardous waste exports from rich industrialized countries to poorer, less-industrialized countries. The name Basel Action Network refers to an international treaty known as the Basel Convention. In 1994, a unique coalition of developing countries, environmental groups and European countries succeeded in achieving the Basel Ban—a decision to end the most abusive forms of hazardous waste trade. Unfortunately, very *powerful governments and business organizations* are still trying to overturn, circumvent or undermine the full implementation of the Basel Ban and in general seek to achieve a "free trade" in toxic wastes.[3]

This description of BAN underscores the organization's belief in the economic basis of global environmental inequality and its commitment to engaging the political economic opportunity structure.

GAIA: Global Anti-Incinerator Alliance, Global Alliance for Incinerator Alternatives

Founded in 2000, GAIA became a global alliance of activists, nongovernmental organizations (NGOs), transnational social movement organizations (TSMOs), scholars, and scientists working for safe, sustainable, and just alternatives to incineration. Based in the Philippines, GAIA has more than five hundred members across seventy-seven countries. Members collaborate across borders to share information and implement coordinated projects around waste prevention, reducing toxics use, recycling, composting, and halting incineration projects. GAIA sees its work as part of a broader movement for social justice and environmental sustainability. A primary focus is ensuring that victories in one region do not translate into additional burdens in another region, or what I call the hyperspatiality of risk. Thus, GAIA campaigns directly against the migration of dirty technologies to the global South. Since the organization is concerned with all forms of waste, it collaborates with other transnational networks fighting pesticide pollution, electronic waste, and health care system wastes.[4]

Global Response (GR)

GR is an international environmental education and action network of activists, students, attorneys, doctors, and educators collaborating to protect human rights and critical ecosystems. This network acts to halt development projects that threaten public health, livelihoods, and critical habitat, primarily in the global South and frequently on indigenous peoples' lands. Its primary mode of activism is mobilizing members in dozens of nations to write letters targeting heads of state, corporate CEOs, or officials at international financial institutions (IFIs) such as the World Bank and the International Monetary Fund (IMF). Comprising more than fifty-five hundred people across ninety-two nations, this network launches campaigns at the request of indigenous peoples and grassroots organizations and seeks to amplify the voices of these local communities to prevent cultural and environmental destruction. The letter-writing approach is a public pressure campaign intended to persuade corporate and governmental officials to make environmentally sound decisions. GR serves as a letter-writing network for communities around the world, often collaborating with TSMO groups like Greenpeace, GAIA, International POPs Elimination Network, BAN, the Rainforest Action Network, and the International Rivers Network. Some of its most famous campaigns have focused on transnational garbage dumping and obsolete pesticides stocks left in global South nations (see chapters 4 and 5).

Greenpeace International

Founded in 1971, Greenpeace International is without question the first organization to successfully raise the global public consciousness about the transnational trade in toxic wastes. Greenpeace representatives were at the Basel Convention meetings from the beginning, and without their presence, that agreement would never have contained the few progressive elements it has. Numerous activists and leaders in other global environmental justice networks got their start as Greenpeace staffers, so the connections among these groups run deep. As Chet Tchozewski, former Greenpeace staffer and current director of Global Green Grants (a foundation that funds grassroots environmental justice activist groups in the global South), put it: "I try to understand the growth of citizens' movements and Greenpeace seems to represent one of the last [twentieth] century's key training grounds for movements that have emerged since then. They have trained and inspired—indirectly through people who have cut their teeth and honed their focus—so many organizations on this. Since Freedom Summer, I don't know if there's a better example of what a movement's done since then aside from Greenpeace."[5]

With thousands of members across the globe and a formal presence in forty nations, it should be no surprise that when ships carrying toxic waste seek friendly ports, Greenpeace is often there to thwart their efforts to dump those chemicals. And like other global environmental justice networks and organizations, its targeting efforts focus on the political economic opportunity structure. As the organization's public literature states, "We exist to expose environmental criminals, and to challenge government and corporations when they fail to live up to their mandate to safeguard our environment and our future."[6]

Health Care Without Harm

HCWH is an international network of more than four hundred organizations across fifty-two nations campaigning to improve public health by reducing the environmental impacts of the health care industry worldwide. Despite being a site of life-giving and life-saving activities, the health care industry is also a major polluter and source of occupational health hazards. Members of the network include hospitals and health care systems, medical professionals, community groups, health-affected constituencies, labor unions, environmental and environmental health organizations, and religious groups. In addition to reducing the pollution generated by the health care industry, one of HCWH's goals is to "promote human rights and environmental justice for communities impacted by the health care industry, while assuring that problems are not displaced from one community or country to another."[7] Again, this attention to the hyperspatiality of risk is a hallmark of transnational environmental justice networks. HCHW collaborates with other networks working on incineration, electronic waste, and pesticides.

International Campaign for Responsible Technology

ICRT is an international solidarity network that promotes corporate and government accountability in the global electronics industry. Its members—environmental and labor activists, attorneys, and scholars from more than fifty nations—are united by a concern for the life cycle impacts of the electronics industry on global health, the environment, and workers' rights.[8]

The ICRT was launched in 2002 by activists and scholars who recognized that the development and sharing of information and critical knowledge across borders were key elements to ensuring a more sustainable and just high-technology industry. The goal was to "ensure that the high-tech industry and governments become accountable to their host communities and people, and that the industry use the best practices to improve health and safety and reduce environmental impacts."[9] This network's membership, leadership, and campaign work overlap with those of other global environmental justice networks and organizations such as BAN, GAIA, Greenpeace, and Global Response. In fact, ICRT's structure was consciously modeled after GAIA's, with leadership balanced between northern and southern nations.[10]

International POPs Elimination Network (IPEN)
Founded in 1998, the International POPs Elimination Network (IPEN) is a global network of more than four hundred public health, environmental, and consumer organizations across more than sixty-five nations, whose goal is to achieve a world free of persistent organic pollutants (POPs). The network encourages and enables NGOs to engage in practices that make measurable contributions to complying with the Stockholm Convention, a UN treaty that provides a framework for reducing and eliminating POPs. As with many other global environmental justice networks, IPEN has a geographically balanced leadership, with codirectors from both northern and southern nations. Their membership and leadership overlap strongly with other global networks, including the Pesticide Action Network, and they work closely with GAIA, Greenpeace, and other groups because their issues of concern are closely interrelated.

Pesticide Action Network (PAN)
The Pesticide Action Network (PAN) is a network of more than six hundred NGOs, institutions, and individuals in over sixty nations collaborating to replace the use of hazardous pesticides with ecologically sound alternatives. Its projects and campaigns are coordinated by five autonomous regional centers: PAN Africa, PAN Asia and the Pacific, PAN Europe, PAN Latin America, and PAN North America. PAN works to empower groups and individuals, particularly farmworkers, women, and people of color, through its publications, conferences, and workshops such as media and Internet and Web skills development.[11]

Taken together, these eight global organizations and networks constitute a formidable presence at international treaty negotiations; within corporate shareholder meetings; in the halls of congresses, parliaments, and city councils; and within local community settings. Even so, they are only a part of the broader global movement for environmental justice. Arguably the most important components of that movement are the domestic local, regional, and national organizations in the various nations and communities in which scores of environmental justice battles rage every day. Those groups provide the frontline participants in the struggles and local legitimacy for TSMOs and networks. Together, the countless grassroots organizations and their collaborating global networks produce and maintain a critical part of the transnational public sphere.

Defining Global Environmental Injustice and Justice

One of the most important tasks that transnational environmental justice movement activists engage in is constructing the problem of environmental injustice. The definition varies from place to place, which makes organizing and collaborating with international partners both a challenge and an opportunity. Paula Palmer is the Global Response campaigns director. She is an activist who has traveled and lived in many parts of the world while working with indigenous and other culturally marginalized groups seeking justice in the face of development pressures, land seizures, and human rights violations: "Indigenous peoples may not define their movements as struggles for environmental protection or environmental justice. Most likely they define their movements as struggles for self-determination and territorial rights. They are fighting for the right to continue managing the natural resources within their traditional territories in the sustainable way they have managed them for eons, and make choices for economic development appropriate to their cultures and environment."[12] In other words, northern activist groups must be attentive to the issue framing that southern activists employ, so as to avoid imposing their frameworks on others or ignoring their struggles entirely.

Sathyu Sarangi is a leading activist in Bhopal, India, who works to secure compensation from the Indian government and the Dow Chemical Corporation for the 1984 Union Carbide chemical disaster in that city.[13] He concurs with Palmer: "I would say, you must know that in India we do not have such compartments of 'environmental movement' and something else which is a 'social movement.' It's all mixed together. There are strong elements of environmental justice throughout every struggle because of the environmental *in*justice that is there, the racism and the double standard are very apparent."[14]

At the Transatlantic Initiative to Promote Environmental Justice, a conference held in Budapest, Hungary, in October 2005, Monique Harden, an attorney with New Orleans–based Advocates for Environmental Human Rights, declared, "Environmental justice means something different in every context." While environmental conflicts generally involve people battling social inequalities and environmental harm, they take many different forms depending on where the struggle is taking place. For activists from Central and Eastern Europe attending that conference, Harden's statement was welcome, particularly since it came from a U.S.-based activist. As much as movement participants around the globe acknowledge that the U.S. branch of the movement gave the cause its strongest push in the early 1980s, there is some criticism that U.S. activists may harbor a privilege and arrogance that arises from the perception that "America is where the movement began." As one global South activist noted, "There is a level of sole ownership, a proprietary nature if you will, to this issue of EJ among U.S. activists."[15]

Echoing that sentiment, Bobby Peek, a world-renowned environmental justice activist and director of the South African environmental justice organization groundWork states:

> American campaigns and activists should not hold a monopoly on environmental justice. The fact that the language of environmental justice was borne out of the Civil Rights movement in the U.S. is great. But I think that environmental justice is going down [happening] in many places in the world, like India and Malaysia, for example; or in terms of challenging Mitsubishi in Latin America; in challenging Shell Oil in Nigeria and the Philippines. Environmental justice is going down daily. Even in Scotland, there are poor folk there in the provinces that are looking at energy and heating services from an environmental justice angle, where you're having poor communities fighting for services, especially for heat in winter. So environmental justice is all around the world.[16]

Another global South activist declared, "Some of the most exciting and innovative actions going on in terms of environmental justice organizing is occurring outside of the United States."[17]

Each of these statements underscores a decentering of the United States as the geographical core of environmental justice movement organizing and expands earlier definitions of environmental inequality to include context-specific, fluid frameworks that could potentially apply to vulnerable people and environments anywhere in the world.

Race, Class, and Nation

Part of the tension concerning the definition of environmental inequality arises from ongoing debates about whether the problem stems primarily from racism or class inequalities.[18] Paralleling debates within the U.S. activist and scholarly communities, transnational environmental justice activists may disagree about the extent to which race and class inequalities play a prominent role in the global distribution of industrial chemicals and hazardous technologies. Romeo Quijano, a leader with the Pesticide Action Network in the Philippines and codirector of IPEN, argues that the primary force causing environmental inequalities is corporate power and the growth imperative: "Basically, the corporate drive for profit is behind it. You understand of course that what makes these big companies thrive is the accumulation of profit, so they need to grow all the time. If they do not have growth in profits, they lose their business survival. Given that situation, it's inevitable that big companies will become bigger, and poorer companies will become poorer, and people in poor nations will be the subjects of environmental discrimination. I will not call it racism but rather environmental discrimination."[19]

Jim Puckett, executive director of BAN, builds on Quijano's sentiments, linking class inequalities to hierarchies between nations: "I think it is more useful to think of it [the international waste trade] in economic terms and not in racial terms. So it's just very clear cut that this dynamic of environmental justice on economic terms is happening globally. People that are poor are getting a disproportionate burden of our toxic crisis."[20]

Heeten Kalan, of groundWork USA, a South African–U.S. TSMO, agrees that economic inequalities are at the root of the problem but also underscores that race matters as well:

> I think that the focus on race and the overemphasis on race is a distraction from class, and that if you keep talking about race in the way that it happens in this country [the United States]—not that it's not important—there's a way in which it can happen that doesn't allow you to talk about class. I think that the powers that be in this country are more than happy for all of us to keep talking about race, but will get really worried when we start talking about race *and* class. This is one way the EJ Movement can actually change the debate, and can actually add something and say this is about race, and its about race and class.[21]

Going further than Kalan, Ann Leonard, of GAIA and HCWH, sees racism as a major factor behind global environmental inequalities. She states that "a driving force for this is simply, I think, international environmental racism. Remember the Lawrence Summers memo? Because there is no other justification I can find for the fact that people are not willing to afford children in other countries the same level of environmental protection that we do in this country."[22]

Finally, Indian activist and journalist Nityanand Jayaraman argued that neither race nor class should be privileged in dissecting the driving forces behind the global dumping and trade of hazardous wastes, because both operate simultaneously:

> The roots of the problem are greed and double standards. In the countries that are dumped upon, the lack of integrity (among government officials), the poor regulatory measures and implementation infrastructure, and a lack of understanding of the dangers of such practices are the main problems. Absolutely race and class discrimination are to blame. There is an unwritten code that justifies development for some at the cost of others (usually, people of color or people from disadvantaged communities or classes). The sentiment that somebody has to pay for development is probably the most damaging and dehumanizing ethic that defines development today.[23]

Jayaramand locates domestic class and political divides as a critical factor in the production of global environmental injustice, something that we observe in many nations. Thus, environmental inequalities are the product of multiple scales and forms of hierarchy that are layered and intersecting.

For Bobby Peek, the solution is to acknowledge both race and class exploitation and push for serious transformations of economic and political systems. He views the path toward environmental justice as one that must move beyond the moderate regulatory policy proposals for reform that many mainstream environmental organizations push for. In Peek's opinion, the movement's focus should not be just on reducing the level of pollution and inequality; the goal must be to reverse these trends entirely, focusing on the political economic opportunity structure:

> It's great that we succeeded in getting environmental justice written into the South African constitution, but the challenge for South Africa is to make that a reality. And making laws that will systematically *unwind and reverse that racism* [from the apartheid era], and give meaning to the principle of environmental justice is what we've got to do. That's going to take a couple of decades of recognizing that black people, poor people, were affected by the types of developments we had in the past and now we have to start *reversing this trend*. Parallel to this, the other important thing is that, as government, as civil society, we need to work on *weakening*

> *corporate influence* on the nation. How do we chip away at corporate power? The other level is getting people to understand that what actually brings us together as a movement, and what can bring us together in terms of challenging *corporate power,* is the issue of environmental justice.[24]

In South Africa, as is the case elsewhere, environmental racism and corporate power cooperate to target people of color, poor people, and their environments and provide privatized benefits to a powerful minority.

As some activists acknowledge, the separation of race and class in the analysis of environmental inequality presumes that racism has an impact on only people of color and does not affect Europeans and white Americans. On the contrary, drawing on the themes of the boomerang effect and racism and class domination as toxicity (see chapter 1), I argue that racism touches all populations, who therefore have a stake in its eradication. Class inequalities operate similarly, creating resentments and tensions that periodically erupt in rebellion and that constantly produce toxic feedback loops (the boomerang effect) that harm both rich and poor.

These questions of inequalities by race, class, and nation are critical to developing the movement's frames that articulate the problems and possible solutions, and they remain unresolved. One way forward might be the connection between environmental justice and human rights, a theme invoked by many activists.

Human Rights

Many activists speak in terms of human rights as encompassing the environmental justice goals of challenging racism, class inequalities, and corporate power. Jim Puckett states: "Human rights is something that is very well understood globally, and even better understood in other parts of the world than in this country [USA], so it is very good that the United Nations has taken the environment on as a human rights issue, with its Human Rights Commission. So we can always talk about human rights and people in other countries totally understand what we're talking about. . . . The compelling nature of the waste trade and toxic trade that hits you in the gut is that it is not only an environmental atrocity but a human rights atrocity."[25]

Movements for environmental justice in the United States are slowly turning to international law and the concept of human rights. This is, in many ways, a significant component of the unfinished business of the civil rights movement, because leaders of that cause were thwarted by the white political establishment from pushing the U.S. movement too forcefully in the direction of human rights and international law, and were instead urged to focus on domestic concerns,[26] and because the civil rights legal framework for addressing "disparate impacts" of environmental policies has thus far yielded little mileage in the U.S. courts.[27] Activists and their supporters are taking note of the various United Nations conventions on racism and environmental rights that constitute a growing body of international law and are being reminded that indigenous communities (particularly in the Americas) have been doing this for quite some time. For example, Monique Harden and Nathalie Walker, attorneys and codirectors of Advocates for Environmental Human Rights, a law center that supports environmental justice organizing, made history in March 2005 when, on behalf of African American residents in Mossville, Louisiana, they filed the first ever human rights petition that seeks reform of the U.S. environmental regulatory system. They filed the petition with the Organization of American States Inter-American Commission on Human Rights because, as they charged, the U.S. regulatory framework fails to protect the human rights of persons in that nation. Harden stated:

> Why do we focus on human rights? Because it is inclusive of and embraces EJ. In the US we only have the Bill of Rights and the Constitution. We don't have Human Rights or a domestic legal framework that recognizes them. We had a forum in Louisiana where the local people defined the remedy for environmental racism in human rights terms. They sent a delegation to the UN to make the case for human rights and EJ based on that work. The US environmental regulatory system violates human rights in at least four ways: 1) the right to life; 2) the right to health (we do not have health-based standards for regulations in the United States); 3) the right to racial equality and non-discrimination and 4) no legally binding requirements for adopting safe alternatives, non-toxics, clean production, and non-discrimination.[28]

Environmental justice activists outside the United States have been deploying the human rights framework for years under the belief that if these norms, practices, and laws have global force, they might prevent another generation of environmental injustices.[29]

Using the overarching banner of human rights, Romeo Quijano links his work in local communities in the Philippines to global environmental justice and progressive politics:

Some of the most significant work I have done in my career is my involvement in campaigns to uphold basic human rights. I became the chairperson of a health and human rights organization here, fighting for the basic rights of people, recognizing their political rights, but also social, economic, and cultural rights, as guaranteed in the United Nations Declaration on Human Rights. Especially in a country like the Philippines where you basically have a grassroots uprising going on because of the inequality of power relations, there are a lot of human rights violations that are happening now. For example, if you look at this new administration, the rate of human rights violations is very high, and that would involve summary executions, arbitrary arrests, and things like that. But of course I also do work on pollutants, specifically against pesticides and other pollutants in the environment. That's why I was involved with the International POPs Elimination Network and the Pesticide Action Network. The link to that work is human rights also. I view the violation of our right to a healthful environment as a violation of our basic right to health. I do it from the perspective of human rights, because all of this is related.[30]

Importantly, transnational environmental justice activists articulate human rights not exclusively in the lexicon of nation-states but also include corporate accountability. As Von Hernandez, a Filipino activist who has worked for GAIA and Greenpeace International, stated: "Even the concept of environmental rights, I think that we just have to keep pushing this concept until they are firmly recognizing this as established international law. Even the concept of *corporate accountability,* which was rarely acknowledged in the Rio Summit papers, but the commitment to a development instrument on corporate accountability recently came about in South Africa at the World Summit for Sustainable Development (WSSD)."[31]

Bridging the 1992 Earth Summit in Rio to the 2002 World Summit for Sustainable Development, Hernandez credits global environmental justice movement networks with successfully pressing the UN to embrace corporate accountability. Challenging, disrupting, and transforming global norms and the accepted common sense about a range of issues are what TSMOs and their networks do best. Equally difficult work involves collaborating with other activists across borders.

Collaboration and Dense Networks

There is considerable overlap among leadership, membership, issue focus, and campaigns among transnational environmental justice movement networks. This is logical from a resource mobilization perspective, because resources for social movements are nearly always scarce. Many of these groups even share office space, providing greater resource efficiencies and opportunities for more intimate and effective collaborations. For example, HCWH and GAIA share an office in Berkeley, California; GR and the funding network Global Greengrants Funds shared an office in Boulder, Colorado for many years; and Greenpeace Southeast Asia and GAIA share an office in Quezon City, the Philippines. Jim Puckett, of BAN and formerly with Greenpeace, commented on the interrelated nature of some of the transnational environmental justice networks: "IPEN became quite a good network and works with BAN and PAN, and in some ways, they're all the same people [laughs]. So all these groups, the ones that are really active on toxics, are going to be active in PAN, they're going to be active in BAN, they're going to be active in IPEN. And now there's GAIA."[32]

Francis Dela Cruz, a Philippines-based activist with Greenpeace Southeast Asia, explained the benefits of having formal and informal ties among activists extending back several years across previous social movements:

> We have a very funky coordination mechanism going around here, we actually share the same building as GAIA. People working in GAIA are like old friends. Like Von Hernandez. . . . I knew Von from his work with Greenpeace and his NGO work before that. But also there was Manny Calonzo, who now works at GAIA and was Von's colleague in the student movement here in the Philippines. They've known each other since '86. And the people working there in GAIA are volunteers of Greenpeace, so I think we are able to share whatever resources, whatever expertise we have quite freely, having that community of people working on this issue. It's great. With GAIA and Greenpeace, and even Health Care Without Harm, we make things happen much more easily. Annie Leonard was telling me this joke about how GAIA and Health Care Without Harm are nursing homes for ex-Greenpeace organizers [laughs]. And because of that I guess it just makes it easier for us to work on campaigns together.[33]

For activists in these networks who have not worked together personally, there is a heavy reliance on technology for much of their collaboration. But this has its limitations, particularly when movement actions often require people to place their reputations, livelihoods, and safety at risk. In this era when most transnational movement coordination is carried out over the Internet and telephone, activists like Jim Puckett see a great need for in-person organizing: "Face-to-face meetings are very important in building trust. When we have these meetings people often say, 'Oh yeah, I know who that person is and we've talked over the phone and e-mail about these issues,' so now people have all met each other and there's this incredible network. This is what happened with IPEN. There are about 150 really dedicated, top-notch activists all over the world in that network, so it's a force that's easily equal to Greenpeace or a big organization like that."[34]

Ann Leonard concurs. When she worked at the Multinationals Resource Center in Washington, D.C., the need for personal contact became apparent:

> Our work involved two main things: one is a lot of work on e-mail, sending out alerts—and the other is contacting people and researching. For example, we found out that the U.S. Trade Development Agency is promoting two new incinerators in the Czech Republic and we immediately whipped off an alert to all the people we know in the Czech Republic. So we ask, what can we do? Should we do a Freedom of Information Act request to get more information for you? That kind of stuff, involving a lot of e-mail. But the other thing that we've found is so essential is you *have* to have face-to-face meetings. You just absolutely have to, because work, especially in other countries, is so dependent on knowing each other and liking and caring for each other. It is so resource intensive, but it pays off like nothing else. There are some people that we worked with for years and years that never quite get in the loop until we bring them to a meeting and they get to meet all these allies that they've been e-mailing with for years and they really run with the issue when they go home.[35]

Each of the transnational environmental justice networks mentioned works to bring participants and members together across international borders for in-person meetings and workshops, often by convening during major global conferences (World Social Forum or UN meetings) or through speaking tours. But transnational collaboration is sometimes easier imagined than practiced.

North-South Networking, Collaboration, and Tension

As noted in chapter 2, one of the major difficulties in building transnational social movement networks is overcoming the legacies of colonial histories and the enduring inequalities that exist between northern and southern activist networks and nations. The physical distance between TSMOs in the North and South is matched only by the social and cultural distance between these organizations. There are often significant gaps between the goals of certain northern TSMOs and local groups in the South. A South Asian activist noted:

> We had seen large groups internationally and in the U.S., which were being dominated by white males. And the way they are structured, they are selling themselves out to the corporations or to World Bank or to both, and to governments. For ex-

ample, making these dirty deals, like say on the issue of allowing toxic waste to be sent to Russia, which was done without consulting local peoples. We knew that we had to be very careful. We don't want our work to be appropriated in a way we didn't want it to be used.[36]

Northern TSMOs "selling out" or betraying the cause of people of color are themes that have emerged with great force since the environmental justice movement's beginnings.[37] These different organizational orientations are quite common and often lead to cultural, organizational, and interpersonal conflicts. The South Asian activist continued:

Yes, I think we have also had to deal with a lot of racism, even among activists. I think that, what we find is that there are people who are well meaning in all other respects, and a lot of the time it works if one points this racism out to them, not so much in confrontation, but making it firm and gentle. That's the kind of policy we have to use. And it has worked in some cases, and it hasn't worked in other cases. And when it doesn't work, in our case, we stop being gentle, to be firm.[38]

Global Response's Paula Palmer is also very much aware of the inequalities between North and South, and her organization works consciously to avoid practicing "environmental colonialism":

It's our policy of offering support to indigenous peoples' struggles *at their request and under their direction*. We don't sit in our office in Boulder, Colorado, and decide what areas of the world need environmental protection and then appeal to indigenous communities in those regions to work with us toward that end. Rather, we respond to requests from indigenous communities that are already engaged in struggles to protect the ecosystems that are their homes. Global Response letter writers support indigenous peoples' political demands as well as their demands for environmental protection. We become allies in their campaigns with the intention of increasing their chances of success. We don't issue statements on these campaigns without their approval.[39]

A number of TSMOs have taken steps to formalize power sharing between North and South, by selecting cochairs or co-coordinators—from both regions and from many representative nations within these regions—to preside over the networks' governance. IPEN and GAIA have done this, as have others. As Annie Leonard, formerly GAIA's northern co-chair, states, "GAIA, in all leadership positions we try to have a balance of people from the North and South, for each workgroup, for each class and course (we offer environmental education courses around the world) and everything; we have a northern and southern co-chair or co-leader."[40]

Other networks have done the same. Paula Palmer noted, "It's important that in our international coalitions, decision making and leadership are shared equitably. For example, the Amazon Alliance, a coalition of indigenous organizations and environmental NGOs in the South and the North, undertook major structural, leadership, and budget changes to ensure equality at all levels. We need to build greater equality into all our international alliances."[41]

There has been a great deal of conflict and mistrust between northern and southern movement groups, and the work of GAIA, IPEN, Global Response, and the Amazon Alliance is in direct response to that history. Among transnational environmental justice activists based in the global North, there has emerged a recognition of their responsibility and accountability regarding the continued transfer of toxic chemicals and other environmental hazards to the global South. This is an acknowledgment of global class and racial privilege.[42] Thus, citizens of the global North have three choices: (1) ignore the problem and their role in the process; (2) gain an awareness of the issue, feel guilty, and do nothing; or (3) educate themselves and employ their privilege strategically and respectfully to address the problem. GAIA and Global Response have consciously employed the third tactic from the beginning. Palmer tells the story of the Global Response's founding and confirms this history:

> The organization's founders were environmental activists in Colorado who thought globally and acted locally. In Boulder, they spearheaded the protests against the nuclear weapons plant—Rocky Flats—created the very successful community recycling organization—Eco-Cycle—and lobbied for the Clean Air Act, the Clean Water Act, the Endangered Species Act, etc. By the early 1990s, they realized that the success of the U.S. environmental movement had an unanticipated and undesired outcome: corporations were packing up and leaving the U.S. in search of locations in developing countries where they would be unburdened by environmental regulations. They were setting up operations in the most politically disenfranchised communities, where local people would find it very difficult to defend themselves and their natural resources. The founders of Global Response felt obligated to start acting *globally* by lending a hand to these communities. They created an organization that would adapt Amnesty International's successful model of citizen letter-writing campaigns to the environmental justice crisis in the developing world.[43]

Thus, Global Response consciously linked environmental justice and human rights in its founding framework.

Kenny Bruno is a former Greenpeace staffer who worked there when the transnational waste trade was just beginning, in the 1980s. He is currently a staff member at Corpwatch, a transnational advocacy group focused on combating the harmful practices of TNCs around the globe. With regard to the questions of unanticipated consequences of the environmental movement's success in the United States, he stated:

> We [at Greenpeace] chose to look at the national borders as a way of stopping the flow of waste, as opposed to state borders or county borders, partly because the environmental movement was successful in pushing for better controls. Someone raised the pressure in the United States—that was part of the pressure that pushed the waste out of the United States. So we actually have the responsibility to make sure that those successes didn't translate into some poorer country somewhere far away getting our waste dumped on them. I believe that was our moral responsibility to look, even at the unintended consequences of some of the successes of the environmental movement here.[44]

All of these TSMOs and networks are clear that the race to the bottom that occurred as a result of stronger regulations in the North is one of the unfortunate legacies of environmental movement "successes" in rich nations. These groups recognize the hyperspatiality of risk as a driver of global environmental inequality. Those TSMOs and networks that are sensitive to this history will have a much greater chance of success in collaborating across national borders. Those groups that discover how to respect the needs of southern and local SMOs, while collaborating with them, will also do well. As Bobby Peek notes, "What's important for people at the grassroots and for the northern NGOs to understand is what drives this campaign, are people suffering, and how do we change that? Another question is how do we speak with the people across these boundaries? I think this is a difficulty in a global campaign, because one needs to make sure that there's a balance between what the communities are wanting on the ground and what big international NGOs are seeking to achieve at another level."[45]

Building on Peek's points, Ravi Agarwal, an international environmental justice activist based in India who works with Toxics Link (New Delhi), GAIA, IPEN, Greenpeace, and HCWH, speaks to the distinctions between global TSMOs like Greenpeace and local, grassroots activism:

> Greenpeace is very good at making issues international. They are very good at working with large international agencies, and they can move things internationally. But it also has a culture of its own, which is based on its own existence. Some local people don't like it when Greenpeace hangs a banner, because we have our own local issues. This is very sensitive in the South. What is the perspective? Whose

eyes are you looking at this from? From one perspective, this is about environment versus development and poverty. We need to lead a healthy life. The issues become very different depending on the scale. Dioxin is not really an issue for the local community. Locally, it's about how a factory changes the use and cleanliness of local water. Then dioxin becomes one aspect of that, but not the key driving factor. This is something in which the local group is best able to do because it understands the local dynamics and what local people really want. Greenpeace has to learn that they cannot do it alone.[46]

Agarwal argues that transnational groups have to better understand their roles in national and local spaces for collaboration to work effectively.

Tensions remain between North and South, but there are also interesting disruptions of the global North-South divide in transnational environmental justice activism, which offers new ways to think about these differences, hierarchies, and hopeful possibilities. Although activists worry about more stringent environmental regulations in the North producing "ecological modernization" in rich nations, coupled with a race to the bottom in poor nations, sometimes we see new political-economic formations occurring. For example, recently the Philippines passed the world's most stringent law banning waste incineration entirely. This event turned upside down most observers' understanding of the global flow of power and the possibilities for social and environmental change (see chapter 4). Another example concerns a South African oil firm that moved to the United States rather than the other way around. South African activist Bobby Peek remembers his conversations with U.S. activists: "We are having a big multinational oil company come out of South Africa, and you're having communities in America saying to us, 'Well, we've got this challenge from your neck of the woods. . . . Help us, what do you know about this?' I think that was an important part of the exchange process in recent years. This created new opportunities for transnational exchange. South Africa has a very rich history of good organizing at a local level against apartheid that the world can learn from, and that America can learn from, in their campaigns."[47]

Peek's colleague Heeten Kalan described these exchanges between the United States and South Africa in ways that revealed many commonalities between communities struggling against environmental inequality, challenging simple North-South distinctions: "The piece also that is very important for me is that South Africans sometimes brushed the U.S. with one stroke. One needs to see that the U.S. itself has its own layers of oppression.

You always here use the term 'North-South exchanges,' right? What I always wanted to do with these exchanges is say it's connecting the South with the South of the North. Because there is a South in the North, right? And who are those people who are the marginalized or the oppressed in the North? Because we have a lot in common with each other and I think that's an important piece for South Africans to see that."[48]

Although South Africa was often viewed as an extreme example of a racial state—given its recent struggles over apartheid—it is also the most industrialized nation in Africa, leading Kalan to refer to his country as "the North of the South in Africa." Each of these moments reminds us that the North-South binary that used to more neatly capture global hierarchies is being destabilized. These disruptions present new openings for local and transnational activists to change policies and challenge state and corporate practices.

Signs of Hope

Although transnational environmental justice coalitions generally emerge out of necessity to battle a particular polluting corporation or government development project, it is the daily solidarity building and exchanges across national borders that provide the necessary foundation for such coalitions and the campaigns they might organize. That was the philosophical rationale for the founding of groundWork, a TSMO with offices in South Africa and the United States. Heeten Kalan of groundWork USA began his environmental justice career by founding the South African Exchange Program on Environmental Justice (SAEPEJ), groundWork's precursor, in the early 1990s as a way of bringing activists across borders to combat the transnational power of states and corporations:

In the beginning, I was traveling from the U.S. to South Africa making connections to the EJ movement here, trying to bridge the roles. I was trying to come to this work with the understanding that governments are talking, multinationals are in constant touch with each other, but the communities that are actually bearing the brunt of all of this, where are they? What are the vehicles that allow them to talk to each other? So how does someone who is being poisoned by Rhone-Polenc in West Virginia talk to someone outside of Johannesburg who is also being poisoned by Rhone-Polenc? What are the vehicles that need to be created there? I had this whole vision that we were a switchboard, we were an exchange, we were looking at how to connect *people on the ground*. And by "exchange" we meant, informational exchange, technical assistance exchange, people-to-people exchange. While the National Toxics Campaign [a now-defunct U.S.-based activist network] was up and running, on that first trip, I took a bunch of samples that we got tested at the lab for a bunch of different toxins from a number of different sites. I remember one of the activists in South Africa commenting on a report I sent them from one of the samples I took. He said that "the fact that I could actually sit with the company and throw this report down—whether that report was valid or not—sent a strong message to them to say you should know that we can access this kind of information within a week." And he said there was power just in that. To be able to say, "We're no longer reliant on you [the polluting company] to give us information and we have our own way to get information" revealed a change in power relationships on the ground in South Africa.[49]

Information is the TSMO's most potent weapon. When sharing information with local activists, TSMOs facilitate challenges to the monopolies on information production that states and corporations often enjoy.[50]

After years of successful organizing by the SAEPEJ, activists decided to create a greater political presence, and Bobby Peek and Linda Ambler established the groundWork organization in South Africa in 1999.[51] Soon after, Peek and Ambler's colleague Heeten Kalan founded groundWork USA in Jamaica Plain, Massachusetts. He relishes the reversal of power relations this signals: "And so, for me it continues this trend of here's this South African NGO that opens up a U.S. office, rather than the other way around."[52]

Once the groundwork of solidarity building and exchanges occurs, transnational campaigns can be more effective. For example, campaigns against locally unwanted land uses (LULUs) such as waste incinerators have provided the flashpoint for collaborations between groundWork and GAIA and hundreds of local, regional, and national organizations around the globe. For instance, in 2003, the Consumers' Association for Penang (CAP, Malaysia) worked with GAIA and many local NGOs to successfully halt a waste incinerator proposed for the town of Semenyih, Selangor.[53] Not only did CAP work with GAIA, but since GAIA is itself a network, CAP was able to access NGOs within GAIA for support. As Mageswari Sangaralingam, a CAP staff member, states: "CAP continued its activities with the Global Anti Incinerator Alliance (GAIA) by networking with other groups in GAIA, exchanging experiences and resources, mobilizing global assistance for local campaigns, signing on petitions against proposed incinerator projects in other areas, writing letters of protest to governments and parties promoting incineration technology and others. Networking of communities and like-minded NGOs involved in similar issues provides significant benefits, which act to increase the chances of success of the cases taken up."[54]

Sangaralingam also notes that it is important that global South NGOs like CAP are actively involved in the leadership of transnational movement networks:

Besides networking with local communities and groups, CAP is also actively involved in international alliances like the Industrial Shrimp Action Network (ISANet), Environmental Law Alliance Worldwide (E-Law), and the Global Anti-Incinerator Alliance (GAIA), etc. CAP holds positions as Steering Committee member or Node in several of these alliances. Being members of these coalitions, we act in solidarity with the other groups, who have similar philosophy, share information and knowledge and support each other in our local efforts. Coordination of campaigns and efforts has been made easier with listservs managed by the Secretariat of the Networks.[55]

Environmental inequality is fundamental to the project of nation building in this late modern era, and activists are keenly attuned to this problem. An overarching concern among southern activists is the drive from within their nations—among governments, industries, and wealthy and middle-class populations—to embrace modernization and globalization. This domestic reality is a core driving factor in the production of global environmental inequalities. Sangaralingam addresses the problem of southern nations' pushing "modernization" projects to more aggressively compete for investment in the global economy and how this dynamic necessarily produces environmental injustices within Malaysia:

Malaysia's shift from an agriculture-based economy to one reliant on manufacturing is reflected in policies aimed at reaching industrialized nation status by the year 2020. The negative impact of this drive to developed nation status is the threat to the environment and health of the people. Since Malaysia's transformation into an industrialized economy, pollution and resource contamination prevail in its river systems, wetlands, marine environments and urban centers. Consumers and communities are now more at risk from dangers of toxic substances. The excessive use of pesticides on crops, paddy fields and even golf courses, coupled with extensive chemical use in manufacturing, pose increasing threats to workers, farmers, and the environment. Emissions from motor vehicles, factories, and power plants, together with open burning of agricultural waste and urban waste, infuse the atmosphere with noxious gases and harmful particulate matters. Untreated waste material and unsanitary disposal practices contribute to environmental health problems in both the urban workplace and in rural areas. Over the years, Malaysia had seen illegal transboundary import of hazardous wastes and also hazardous industries. Due to loopholes in our legislation and probably ignorance or greed of decision makers, foreign-owned hazardous industries have found their way onto Malaysian soil. These manufacturers, who find it difficult to set up a factory in their own country due to public protest or strict laws, move to developing nations with their harmful technologies.[56]

Given all of these forces converging to produce global and local environmental inequalities, it is a wonder that activists are ever successful in their campaigns and a greater wonder that they are able to continue their work with hope.

Visions of Global Environmental Justice

Echoing Mageswari Sangaralingam's statement, Indian activist and journalist Nityanand Jayaraman spoke to the brutality of the political economic opportunity structure and global environmental injustice, with an eye to both its domestic and transnational causes: "Given that most of the governments are disproportionately influenced by industry, it is an uphill battle to achieve success. In countries like India, one has to counter the argument that hazardous wastes are required to feed the waste recycling industry and hence a priority for national development! Exacerbating these challenges is the role played by the U.S.—which can be characterized as the single, most powerful environmental villain. In fact, the policies of the U.S. government are so openly racist that one becomes frustrated even before beginning a campaign."[57]

Activists interviewed for this study articulated the roots of the problem of global environmental injustice in terms of race, class, national, and international inequalities that are practiced primarily by governments and corporations, and secondarily by some northern environmental groups. Although there is a range of opinions among activists about the relative weight that should be accorded to racism versus class inequalities, all agree that environmental inequalities are most immediately a result of political and economic institutional power. This common sense of the problem is what social movement theorists call a diagnostic frame—the source of their grievances and therefore the target of their activist energy.[58] Producing a diagnostic frame is the first step toward mobilizing to achieve a solution, a goal, and ultimately a vision of a better world. Jayaraman discussed his vision for what role governments and corporations should play in relation to communities and ecosystems:

> The role of governments? Very simple. Protect public health and environment. Acknowledge and respect the community's relationship and command over natural resources. Assist in the development of progress schemes that utilize local natural and human resources and are geared toward meeting the basic needs of the community. The government needs to remain unbiased and recognize that corporations do not have the same rights as people. As trustees of public resources, they should refrain from handing over natural resources to private entities at the expense of the needs of a community. Most important, governments should have the vision and the ability to safeguard our resources for the generations to come. Industry's role should be to cater to the real and basic needs of human communities before addressing the luxuries. They should charge for their services and should fulfill their roles without exposing their workers or the environment or the community to undue hardships.[59]

This is a vision of global social change that entails building and supporting sustainable and equitable communities and challenging industry and government to work for the good of the disenfranchised majority. These transnational environmental justice activists view the global political economy as shifting risks and hazards from North to South, from rich nations to poor communities between and within nations, and from racially privileged communities to racially despised communities. They are radical in their approach, which means they challenge the sources of power in the global political economy: governments, TNCs, IFIs, environmental groups, and racist and classist ideologies. These forces—these targets—constitute the political economic opportunity structure that activists seek to destabilize, disrupt, and open up in the hope of creating a more just transnational political space.

In the following chapters, we consider the work of activists who are collaborating across social and geographic borders to challenge historical and ongoing environmental injustices, thereby encouraging the embrace of a more open and influential global civil society.

4

The Global Village Dump: Trashing the Planet

Since human societies first produced waste streams, we have buried our trash, and since the discovery of fire, we have burned our waste. Dumping and incineration are the oldest methods of waste management known to humankind, and they remain the most prevalent methods in use. Those of us in the global North produce more waste than the rest of the world, and we have become more sophisticated in our efforts to encourage other communities and nations to host and manage our waste for us. Wealthy communities in the global South engage in similar practices to find the path of least resistance in waste management. At the same time, local, national, and transnational efforts to oppose the practice of transboundary (across nation, race, and class borders) garbage dumping and incinerator sitings (i.e., technology transfer) have grown in strength and influence as well. In this chapter, I consider the problem of global garbage dumping and the contours, reach, and power of transnational environmental justice networks to confront the timeless practice of dumping and burning waste in politically and economically vulnerable communities. I integrate Charles Mills's theory of black trash into my framework to understand how environmental inequality has become a de facto global policy of states and corporations dealing with modern waste streams.

Charles W. Mills draws on philosophy and historical texts to connect white racism to a psychological, cultural, and legal framework that links images of people of color (specifically people of African descent) with barbarism, filth, dirt, and pollution.[1] According to Mills, African peoples themselves are viewed by many whites as a form of pollution, hence making it that much easier to contain industrial waste and factory pollution in their nations and segregated neighborhoods. The link between non-European peoples and symbols associated with nature, such as danger, disease, filth, and the primitive and savage, is common throughout European history, literature, and within contemporary politics in the global North, whether one is speaking of Africans, African Americans, indigenous peoples, Asians, Latin Americans, or the Roma of Europe.[2] But these ideologies are also present within non-European societies where despised others exist—whether they are indigenous peoples in Mexico or Dalits or Untouchables in India who are viewed as a form of contamination. Thus, it is a globally ubiquitous racial construction. Like Mills, environmental philosopher Robert Higgins identifies the cultural sources of meanings of racial and social pollution, in that minority environments are seen as "appropriately polluted" spaces. He argues that racial environmental inequities are a product of the social deployment of power in the "mastery of nature." Racial segregation at work and at home, insofar as it generates perceptions of populations as pollutants, facilitates the environmental burden placed on those communities because "environmental pollution is fittingly relegated to 'socially polluted spaces.'"[3] Immigrants, indigenous populations, and peoples of color are viewed by many policymakers, politicians, and ecologists as a source of environmental contamination, so why not place noxious facilities and toxic waste in the spaces these populations occupy or relegate these groups to spaces where environmental quality is low and undesirable? Such practices would not violate any cultural or political ethic if the previous point holds true. Mills and Higgins provide a framework for understanding why environmental racism is so prevalent and culturally logical.

Economic or class hierarchies between and within nations function to reinforce domestic and global racial inequalities and vice versa. Ideological justifications for the growing class divide, which come from political leaders, journalists, and academics, tend to implicitly blame the poor and laud the rich for their divergent social positions.[4] And as others have argued, if the domination of nature is part of a broader project of class domination inherent in capitalism, then environmental inequality is the logical result. Thus, in a global political economy, race, class, and national hierarchies may be dynamic, but they are always at work and serve to maintain an exploitative relationship to nature as well.

Thus, it is not difficult to understand why entire regions of the globe (notably much of the global South and many areas within regions of the global North) have been "redlined" in that these places are largely viewed as available for the transfer and dumping of the cast-offs and negative externalities of northern production and consumption practices. This process of global environmental inequality is made possible not only by ideologies and discourses that produce and intensify social distance between rich and poor and North and the South,[5] but also by the practices of domestic spatial and residential segregation and the furthering of international and geographic inequalities through institutions like the World Bank and the International Monetary Fund, and by global free trade agreements and the corporations that benefit from them. These ideologies, institutions, and practices all serve to produce an out-of-sight, out-of-mind mentality for groups in the North that can ignore the reality of domestic and global environmental inequality by transboundary waste transfers, while their day-to-day living routines produce these very terrors.

The dumping of trash, garbage, or municipal solid waste remains the most common form of environmental inequality. In the late 1970s and early 1980s, low-income and working-class communities across the United States rose up against the disproportionate dumping of such waste in their boundaries, highlighting the problem of domestic environmental injustice.[6] Since the late 1980s, environmental inequality associated with siting garbage dumps and exports of hazardous waste has become a routine transnational practice. The rise of global environmental justice movement networks is a grassroots response to that reality.

The Rich Get Fat, the Poor Get Trashed: Overconsumption and the Mass Production of Garbage

Dumps and Landfills

The volume of municipal solid waste generated in the United States between 1998 to 2001 grew by 66.6 million tons, or 20 percent, to a total of 409 million tons per year. Of that total, government and industry landfill about 60 percent, incinerate about 7 percent, and recycling or compost 33 percent. The number of landfills declines every year as they reach maximum capacity or are shut down as a result of public concerns over pollution.[7] The point is that the United States produces more and more waste each year and the issue has nearly always been hotly politicized. The volume of trash that U.S. residents generate has doubled in the past forty years. What is further cause for concern is how the United States compares with the rest of the world. The United States has only 5 percent of the world's population but generates 19 percent of its wastes. The United States uses 20 percent of the world's metals, 24 percent of the energy, and 25 percent of the fossil fuels. At 31.5 pounds, the weekly waste output of the average U.S. citizen is the highest in the world. Yet among the twenty most industrialized nations, the United States is fifteenth in paper recycling and nineteenth in glass recycling. Nearly 90 percent of plastic and 50 percent of the paper generated in the United States ends up in landfills. Mexico recycles more glass than the United States does.[8] However, recycling is no salvation. First, the very essence of market economies is predicated on continuous growth, and although there has been a marked increase in materials recycling in the United States and United Kingdom since 1960, this has been far outpaced by the growth in consumption and waste production.[9] Second, much of the paper, plastic, metal, and electronic goods collected for recycling in the United States is actually exported to the global South, to be dumped or used in various remanufacturing processes that are typically more hazardous to workers, communities, and ecosystems than would be allowed in the United States.[10]

Although the United States may be the worst violator of global environmental ethics, it is only one part of the much larger pattern of overconsumption and production within global North communities. On average, residents of the richest countries throw away up to 1,763 pounds of waste each year, compared to less than 440 pounds in the poorest countries.[11] Speaking to the same issue in the United Kingdom, the progressive environmental group Friends of the Earth makes it clear that the disparity between the level of consumption in the North versus the South is unsustainable: "We are a nation addicted to chucking stuff out. How many worlds do we need? If everyone in the world was as wasteful as we are in the UK we would need 8 worlds to keep going."[12] This figure is particularly problematic given the projected estimates that England's municipal solid waste volume will double by 2020.[13] And that figure is restricted to municipal solid waste, or garbage. Perhaps more worrisome for Britons (and, by implication, the global South, which might eventually become hosts of their waste) is that Britain's capacity for treatment and storage of hazardous chemical wastes is decreasing rapidly, raising fears of future illegal dumping schemes and exportation. Like the landfill crisis in the United States, this dilemma in the United Kingdom has little to do with actual available space for dumping; rather, it stems from the political reality of increasingly stringent regulatory requirements for waste management and processing firms demanded by northern citizens and environmentalists.[14]

One recent study in England concluded that rubbish is dumped illegally in that nation every thirty-five seconds, costing authorities in London 100 pounds sterling (US$191) per minute to clean up, prompting the Department of Environment Food and Rural Affairs, the Environment Agency, and the Local Government Association to develop "Flycapture," a database set up to monitor and enforce the law against "fly tipping," that is, illegal waste dumping.[15]

This seemingly extraordinary production of waste is a normal part of the so-called development process practiced by market economies. As nations become more heavily industrialized and committed to the global capitalist economy, the production of waste and environmentally harmful substances increases. As a British Broadcasting Corporation report declared, "The more 'developed' we get, the more we throw away."[16] Consider the town of Bray, Ireland, where some 200,000 tons of trash were buried over the course of several generations at a municipal landfill. The dump was closed in the 1980s but is still quite open: it was built on the edge of the ocean and is now falling into the water. Ireland has become known as the Celtic Tiger economy: with a major influx of investment over the past several years, economic development has increased rapidly, creating both a fiscal boom for some and a vast increase in the volume of garbage.[17]

No society produces zero waste. In fact, many global South nations produce waste that is poorly managed to begin with, so the import of garbage from other nations only adds international insult to domestic injury. Consider the capital city of Baku, Azerbaijan, which has an immense dump site sprawling over thirty hectares (seventy-four acres) of land. Like landfills all over the world, this site offers both an opportunity for scavengers (gold chains, leather jackets, and scrap metal that are recovered and sold provide a living for around four hundred people) and dangers for residents (risk of salmonella, dysentery, botulism, or cholera). There was once a recycling plant in the area, but it has stood idle since the breakup of the Soviet Union.[18] The practice of waste scavenging and reuse is almost invariably much more widespread in the global South than in the North, not because people in the South are more environmentally conscious but because the grip of poverty makes reuse an economic necessity.

Although there is no doubt that the transnational trade and illegal dumping of waste exacerbate public health and environmental harms in other nations, it is not my purpose to try to prove that point; rather, the goal here is to underscore that these practices reflect social, political, cultural, and economic hierarchies between and within nations and communities of the global North and South and that international waste dumping is not only an environmental injustice but an extension of unequal relations and hierarchical practices that characterize the everyday, routine functioning of the global political economy. As Xing Demao, director of the Bureau for Inspection for Shandong Province in China, asked, "Why do some countries strictly control export of their advanced technology and equipment but openly permit the export of harmful waste they produced?"[19] Demao was responding to reports that U.S. firms had illegally hidden garbage inside otherwise legal shipments of waste paper bound for recycling facilities in China. Soiled diapers were found in one shipment, "creating a stench so bad that made it difficult for officials to even approach, much less open and inspect it."[20] In the wake of this incident, the words "foreign garbage" became a nationalist rallying cry in the media and streets of Chinese cities, aimed at what is believed to be a widespread American disrespect for China.[21]

In U.S.-occupied Iraq, scavenging a growing rubbish dump—which many people in the middle-class residential district of al-Jihad assume comes from coalition civilian or military sources—provides a supplementary income for some lower-class Iraqis. However, other local residents are appalled by the refuse, claiming that it is a health hazard and is culturally offensive. "When the CPA [Coalition Provincial Authority] dumps rubbish in the middle of the neighborhood, it hurts Iraqis' honor," said Fowzi Hussein. Referring to the pornographic magazines that are sometimes found in the waste piles, Nihad, a middle-aged man, said, "It is immoral, uncivilized, and against our religion." Others threaten that if the CPA does not stop dumping rubbish, residents will begin conducting resistance. "If the Americans continue, a terrible battle will ensue," said taxi driver Khalil Mahir.[22] The al-Jihad dump is but a symbol of U.S. domination of the Iraqi people and their environment. While U.S. citizens dump on these neighborhoods, the United States is, at the same time, extracting a fortune in natural wealth contained in the countless barrels of oil being siphoned off, leaving Iraq polluted and impoverished. Combined with the U.S. military invasion that began in 2003, this is a particularly brutal example of environmental racism.

Thus, although all societies produce and dispose of waste, some produce much more than others and have insisted on exporting it to weaker nations, where the people are devalued and comparatively disempowered. In the United States, where the environmental justice movement began in the 1980s, this pattern holds true. Landfills and illegal waste dumps do not checker the landscape randomly; more often than not, they are placed in areas that are home to low-income communities populated with people of color.[23] This domestic pattern of environmental racism and inequality extends to other nations and is marked not only by dumping waste in landfills near vulnerable communities but also by placing waste incinerators in those spaces as well.

Incineration

When dumping waste in a landfill becomes impractical or politically unpopular, societies have typically burned that waste. From small pits to large machines called incinerators, human societies have treated waste in this fashion for as long as anyone can remember. Waste incineration technology has grown more sophisticated in the past century and, like landfills, tends to find its location in communities that are culturally, economically, and politically marginalized. A 1990 Greenpeace report on incinerator siting practices concluded that communities with existing incinerators had populations of people of color 89 percent higher than the national average, and communities with proposed incinerators had populations of people of color 60 percent higher than the national average. That is, the darker the community, the greater the likelihood of finding an incinerator located there.[24] In 1997, 15 percent of the U.S. nonwhite population lived within two miles of a medical waste incinerator, while only 9 percent of the Anglo or European American population did.[25] This was no coincidence and not the result of the invisible hand of capitalism. This pattern of discriminatory facility siting was documented as a national phenomenon in the United Church of Christ Commission on Racial Justice's 1987 report, *Toxic Wastes and Race in the United States*. That year, an earlier report came to the public's attention, revealing the otherwise hidden face of power behind these trends. In a 1984 report, which Cerrell Associates produced for the California Waste Management Board, the author wrote: "All socioeconomic groupings tend to resent the nearby siting of major (waste disposal) facilities, but the middle and upper socioeconomic strata possess better resources to effectuate their opposition. . . . Middle and higher socioeconomic strata neighborhoods should not fall at least within (five miles) of the proposed site." The study indicated that incinerators placed in rural communities that were politically conservative, above middle age, Catholic, and poorly educated were less likely to encounter resistance because these populations are more accepting of what are otherwise viewed as locally unwanted land uses (LULUs) than are other communities.[26] This was a "how-to guide for state officials looking for politically vulnerable communities in which to place incinerators."[27] The location of incinerators in the United States was a clear issue of environmental inequality and environmental racism, as incinerator proposals increased dramatically during the 1980s and 1990s, a result of environmental laws that placed greater restrictions on waste management facilities than ever before. In 1980, the United States incinerated only 1.8 percent of its solid waste, but by 1990 there was an eightfold increase to 15.2 percent.[28]

Modern incineration technology was developed in 1885, yet remains intensely unpopular with many communities where these facilities are located. According to Paul Connett, a scientific authority on incineration and the leading opponent of waste incineration, "incineration is the most unpopular technology since nuclear power."[29] The antitoxics movement and environmental justice movement in the United States rose up during an incinerator boom in the 1980s and 1990s to meet this threat to local communities and challenged numerous incinerator proposals, shutting down existing burners across the nation.[30] Because of the overwhelming focus and presence of the environmental justice and antitoxics movements on this issue, between 1985 and 1998, plans to construct more than three hundred garbage incinerators were defeated or put on hold in the United States. As table 4.1 reveals, there was a steady rise in the construction of

Table 4.1
Municipal waste incinerators in the United States

Year	Number of municipal waste incinerators
1965	18
1970	25
1975	45
1980	77
1985	119
1990	186
1995	142
2000	116
2002	112

Source: Tangri 2003, 66.

Table 4.2
Decline of medical waste incineration in the United States

Year	Number of medical waste incinerators
1988	6,200
1994	5,000
1997	2,373
2003	115

Source: Tangri 2003

municipal waste incinerators in the United States until 1990, when they began to be dismantled.[31] We also observe the same general trend with medical waste incinerators, as reported in table 4.2.[32]

While it may come as no surprise that environmental justice activists would take credit for this decline in the prevalence of incineration in the United States, the federal government has agreed with this assessment. According to the U.S. Department of Energy in 1997: "There have been increasing challenges to interstate waste movement. . . . With increasing awareness and protest by communities, the governments have been forced to involve them in the decision-making process. This sometimes means having to leave the waste management option to the communities themselves. People are increasingly opting for recycling and composting of waste, and out of WTE [waste-to-energy]."[33]

Unfortunately, the success of this movement activity also gave industry an incentive to export waste beyond the United States. As incinerators continue to suffer defeat in the United States and Western Europe, there has been a proliferation of proposals for such technology in the global South and Eastern Europe, the result of the industry's geographic shift to follow the path of least resistance. As South African environmental justice activist Bobby Peek stated, "In South Africa generally, there has been an increase in companies wanting to burn hazardous wastes as a fuel. For us, that is an incineration debate, and that is a waste debate."[34] The grassroots anti-incinerator sentiment has, in turn, blossomed into a global phenomenon, the most visible manifestation of which is the emergence of the Global Anti-Incinerator Alliance (GAIA).

Waste incineration has been proven to yield poor economic results for communities, working conditions are often hazardous, and they generally emit the following toxins: persistent organic pollutants (POPs) such as dioxins, furans, and polychlorinated biphenyls (PCBs); heavy metals such as lead, cadmium, methyl mercury, chromium, and arsenic; and gases such as sulfur dioxide and nitrogen oxide.[35] As GAIA activist Ann Leonard states, transnational waste incinerator sitings are illogical for host communities because "it makes absolutely no sense to pay someone from another country to set up a machine that will burn your resources that you could be reusing and generating income from."[36]

Grassroots community activist groups like People for Community Recovery (Chicago), Mothers of East Los Angeles, and national and international organizations like Greenpeace and GAIA have worked hard not only to shut down existing incinerators and halt construction of planned burners; they seek to challenge the legitimacy of incineration technology itself. Although they have had success in many corners of the world, the convergence of industrialists pushing this technology, elected officials receiving favors from this industry, and the growing volume of waste being produced in the North and increasingly the South, has produced a gargantuan challenge for the global anti-incinerator movement.

In the rest of this chapter, I consider several cases of postconsumer waste dumping and waste technology transfers from North to South. Each case reveals a distinct aspect of the problem of global environmental inequality and of the movement for environmental justice and human rights that has responded to this problem. And while North to South dumping is evident all over the planet, we must also consider the fact that elites and the middle classes within southern nations also contribute a great deal to the problem because they generate waste domestically, which is dumped in poor, indigenous, and ethnic minority communities within their own nations. Southern elites also engage international and national political-economic systems to facilitate this dynamic. Thus, environmental inequality is the result of hierarchies across race, class, and spatial boundaries that span the international, national, and local scales.[37]

Operation Return to Sender

Philadelphia Becomes a Symbol of America's Trash Crisis

In 1986, Wilson Goode, the mayor of Philadelphia, had a major political headache on his hands.[38] The city had countless tons of incinerator ash and no landfill in which to place it. Beginning in the late 1970s, the city burned 40 percent of its municipal solid waste in two incinerators and buried it in the Kinsley landfill across the Delaware River, in New Jersey. Then in 1984, New Jersey closed that landfill to Philadelphia's ash. This was a time when the U.S. antitoxics and environmental justice movements were gaining ground, and landfills and incinerators were some of the main targets of campaigns to shut down LULU's. It was also a time when the average U.S. resident was increasing the volume of waste he or she produced, so there were myriad pressures contributing to a solid waste crisis in Philadelphia. The city needed to export the ash somewhere because, as a result of the heightened politics of waste, there was no space for it locally.

In 1986 the city contracted with a local paving company, Joseph Paolino & Sons, to get rid of fifteen thousand tons of municipal incinerator ash. Paolino & Sons turned the waste over to Amalgamated Shipping, a company headquartered in the Bahamas, which owned a ship named the *Khian Sea,* a 486-foot, seventeen-year-old rusty vessel registered in Liberia. With its cargo of ash onboard, the ship left Philadelphia on September 5, 1986, headed for the Bahamas. That moment set off a sixteen-year journey that would see this ash travel around the world, from port to port, sparking one of the most infamous controversies in the long saga of the international waste trade. This was the first major documented case of transnational waste dumping by the United States and is believed to have been the primary driving force behind the development of the Basel Convention regulating the transboundary movement of hazardous wastes. But the untold story of the *Khian Sea* is that it was an unbridled example of international environmental racism and a case that became an early training ground for the movement for global environmental justice.[39]

Shortly after the *Khian Sea* left Philadelphia, Bahamian officials concerned about the potentially toxic content of its cargo turned the ship away. It then traveled to the Dominican Republic, Honduras, and the West African nation of Guinea-Bissau, all of which refused the waste.[40] Then the ship tried, and was refused entry to, ports in Puerto Rico, Bermuda, and the Netherlands Antilles. Eventually, in December 1987, the *Khian Sea* landed at Gonaives, Haiti, one hundred miles north of the capital city of Port-au-Prince. Upon arrival, under an agreement with the military regime of Lieutenant General Henry Namphy, the crew unloaded 3,700 tons of ash and pulverized glass on the beach, where it was soon discovered that the cargo was falsely labeled "soil fertilizer."[41] When local environmentalists and authorities became suspicious of the ash, Robert Dowd, an Amalgamated Shipping staff representative already aboard the ship, stepped ashore and pronounced the waste safe and clean. He stated, "We are not sick, there is no smell and there are no flies." Sensing that the local residents were unconvinced, he did something no one expected: he bent down, scooped up a handful of the ash, and ate it: "See, it's edible. No danger."[42] The Haitian government was neither amused nor persuaded and ordered the ash removed immediately. But the *Khian Sea* had departed under cover of darkness and left the ash on the beach. At that point, 10,000 tons of waste remained aboard the ship. In order to dispose of the rest of the cargo, the crew attempted to secure agreements with Senegal, Cape Verde, Sri Lanka, Indonesia, and the Philippines. All refused them. The remainder of the cargo was never fully accounted for, but at least 10,000 tons later disappeared between the Suez Canal and Singapore. Officials surmised that it had been dumped somewhere in the Indian Ocean.

Every nation and every port where the *Khian Sea* attempted to dump the ash was a nation with an extremely low gross domestic product (GDP), high rate of poverty, and majority populations of people of color. The Environmental Research Foundation's Peter Montague called this an "act of international environmental injustice," and Greenpeace is said to have referred to the *Khian Sea*'s dump at Gonaives as "one of the most blatant and long-standing examples of environmental racism and injustice worldwide."[43] At the time, "Haiti [was] the poorest country in the hemisphere, with a GDP in 1990 of about $2.4 billion and average per capita income of $380. The city of Philadelphia has a budget of $2.6 billion, and per-capita income is $25,055, according to the Statistical Abstract of the U.S. In comparison to Haiti, Philadelphia is fabulously wealthy."[44]

Activists in the United States and Haiti united to publicize the case and pressure the U.S. government to act. In response, the U.S. Environmental Protection Agency (EPA) and the United Nations Development Program sent representatives to the site and recommended the construction of a permanent containment facility in Haiti in which to hold the ash.[45] Due to growing concerns over public and environmental health risks associated with the ash dump, Haitian authorities later removed the waste and sealed it in concrete on a hillside a few miles away from the original dump site.

In December 1991, the military-led government was removed in elections that placed progressive Roman Catholic priest Jean-Bertrand Aristide in the president's office. This development changed the political opportunity structure for activists seeking to clean up the ash and facilitate a stronger civil society presence in Haiti. A short while later, when Aristide was exiled from Haiti in an attempted coup, the U.S. military invaded the country to restore his government. Sensing an opportunity, Greenpeace sent a request to the armed forces that they consider bringing the incinerator ash back to the United States after completing the mission in Haiti. Thinking back on that moment, Greenpeace activist Kenny Bruno remarked, "I think they giggled," because the proposal was not taken seriously.[46] But the Haitian government and Haitian activists were determined to make a change. Evans Paul, the mayor of Port-au-Prince, traveled to Washington, D.C., in May 1992 to seek congressional and public support for the removal of the ash. "Basically, we believe that the people who put it there should remove it," Paul stated. "And our people should benefit from any penalties."[47] Likewise, Haitian-American organizations such as the Haiti Communications Project led fact-finding visits to Gonaives in December 1991 and again in March 1992. After the first visit, a representative of the group returned to the United States with five hundred envelopes, half of them addressed to Philadelphia mayor Wilson Goode and the other half addressed to William K. Reilly, chief administrator of the U.S. EPA. A sample of the incinerator ash was placed in every envelope, which carried the label: "Contains Philadelphia Waste. Return to Sender. Delivered three years ago. Mislabeled as fertilizer."[48] Despite these creative and dramatic efforts, no progress was made on the case over the next several years.

Politicizing Waste: Activists Mobilize "Project Return to Sender"

The *Khian Sea* owners were eventually tried and convicted on federal perjury charges and served time in prison. The Paolino & Sons company was the subject of a criminal investigation, but none of its officials or employees was ever prosecuted. However, environmentalists did not keep quiet, continually pressing the City of Philadelphia to take responsibility for its incinerator ash. The city resisted steadfastly. Then, in 1996, a seemingly unrelated event elsewhere opened up a point of access for activists. That year, New York City created the Trade Waste Commission whose charge was to "clean up" the city's waste hauling industry, which had nearly always experienced an undue influence of corrupt officials and organized crime.

Soon afterward, the commission received an application from Eastern Environmental Services, a New Jersey–based garbage hauling company. The application was for a permit to collect and dump waste from businesses in New York City. In a routine procedure, the commission conducted a background check on Eastern Environmental and found that one of its executives, Louis Paolino, had previously managed the Paolino & Sons company, which had been the subject of intense media attention and criminal investigations concerning the Philadelphia ash dumped in Haiti. The commission was concerned about this unsavory history and, during negotiations over whether Eastern would be granted a license to haul trash in New York, city officials proposed that the company clean up the ash in Haiti as a sign of its good faith. Given that the profits from any contract with the city would provide the company a revenue windfall, Eastern agreed to pay $100,000 toward the removal of the ash from Gonaives. The total cost of the removal and return of the waste was estimated at between $150,000 and $250,000, so the remaining funds would have to come from elsewhere. The deal between the commission and Eastern stated that the remainder of the funds would have to be secured by May 1998 by other parties, or the company would be relieved of its obligations and allowed to secure the permit.

This announcement sparked great interest among activists, who seized on it as a new opportunity to bring the ash home to Philadelphia. Kenny Bruno stated, "This is a very positive and laudable initiative. New York City didn't have anything to do with the original scandal, but it's attempting to make the company show some good faith and clean up a past wrong." Edward T. Ferguson, chairman of the Trade Waste Commission, sent a letter to Mayor Edward G. Rendell of Philadelphia (Wilson Goode's successor), stating, "We hope you agree that it would be appropriate for the City of Philadelphia to commit the relatively modest sum of $50,000 toward the removal of its incinerator ash from an illegal dump site in an impoverished foreign country." The mayor's office rejected the request on the grounds that the responsibility rested with the waste contractors, not Philadelphia. Activists were outraged, as were Haitian officials and Haitian American leaders. Ehrl D. Lafontant, the executive director of the Haiti Communications Project (based in Cambridge, Massachusetts), stressed that Philadelphia should acknowledge its role in the scandal and take part in the solution. "Whether it was a contractor or a subcontractor who dumped it," he said, "Philadelphia has some responsibility for this gross act of environmental injustice."[49] It is instructive to note that Lafontant framed this case in the language of environmental (in)justice, in order to signal his organization's view that this was the most effective articulation of the problem and reflecting the reality that his strongest political networks on this case were with environmental organizations.

Activists in Haiti and the United States used this opportunity to reignite the campaign to return the ash. In 1998, several international and national social movement organizations in the United States and Haiti combined forces to create Project Return to Sender to pressure U.S. government officials in Philadelphia and Washington to repatriate the waste. Essential Action, a Washington, D.C.–based organization, was the center of the network.[50] In one of its action alerts, Essential Action cited then Vice President Al Gore's best-selling book *Earth in the Balance,* in which he decried the transnational waste trade as "disquieting" and "unthinkable." Essential Action then called on citizens around the world to write to Gore to act on his concerns about the waste trade and bring the ash home to the United States from Haiti:

> Now that Al Gore serves as Vice President of the United States, the country responsible for this atrocity, we ask Mr. Gore to put actions behind his words and rectify the damage.
>
> We need to obtain another $200,000 fast. So far, Philadelphia is refusing to help. We believe the government of the United States bears responsibility for repatriation of the waste because: 1) The waste was generated in the USA; 2) The waste in question has now become a foreign affairs issue violating the spirit of the Basel Convention which both Haiti and the USA has signed [although not ratified] . . . the USA is an OECD country and Haiti is a non-OECD country [and the Basel Ban disallows waste trading between such countries]; and 3) the waste's presence on Haitian soil also violates Principle 21 of the Stockholm Declaration.[51]

Citing the Basel Convention and the Stockholm Declaration is a tactic that international environmental movement groups practice because it raises the visibility of the struggle beyond a conflict between two states and also raises the stakes for those accused of committing crimes or violations of such policies. The action alert also declared that the

> City of Philadelphia should contribute the remaining funds to finally solve this 10–year environmental crime because: 1) The waste was generated in Philadelphia, PA, USA; 2) Philadelphia never paid for the original disposal of the waste. The City refused to pay the waste hauler because the waste was disposed of improperly, saving the City budget over $600,000 according to Greenpeace; and 3) the City of Philadelphia had a $130 million budget surplus last year. A contribution of $200,000 is only 0.008 percent of Philadelphia's annual budget, a small price to pay to remove this toxic threat from Haiti and win a victory for international environmental justice.[52]

This action alert urged citizens to write to Al Gore, the mayor of Philadelphia, and the United States Agency for International Development (USAID) to contribute funds to repatriate the waste. Thus, it was clear that while activists seemed to be engaging a state-centric model of the political opportunity structure, they were also doing so at a number of scales, involving local, national, and international organizations in the United States and Haiti, to pressure governments at multiple levels. Moreover, in this case, the corporate stakeholders were viewed as both perpetrators and problem solvers, since Amalgamated Shipping and Paolino & Sons were jointly responsible for the waste being dumped in Haiti, while Eastern Environmental Services was contributing to the ash removal project. So in an interesting twist on the political-economic opportunity structure model, environmentalists found themselves opposing what one arm of this corporate network had done while working with another arm to clean up the ash. Hence, both the state and industry are central to environmental justice struggles and solutions. In addition to demanding government action on this issue, Greenpeace activist Kenny Bruno punctuated this point in a faxed letter to Project Return to Sender member organization Global Response. He wrote: "To reiterate the points I made on the phone about why this is an appropriate Global Response Action . . . Lesson in *corporate accountability:* It is a wonderful lesson that Eastern Services—even though it is not the original company which dumped the waste, and even though it was only one of several actors at fault, and even though 10 years have passed—will still have to pay the price for the misdeed of its CEO if it wants to expand its business."[53]

In March and April 1998, a group of U.S. activists with Project Return to Sender traveled to Port-au-Prince to meet with partner organizations there. They also held a number of joint protest actions in front of the U.S. embassy there to demand the repatriation of the ash. Witness for Peace and Voices for Haiti demonstrated on March 13 and April 24 that year and visited the ash dump site. "My garbage was burned in Philadelphia. Now it's in Gonaives," said Ray Torres, a member of the U.S. activist delegation. "It's a moral outrage."[54] Activists delivered a bag of the ash to staff members at the U.S. embassy in a symbolic gesture of "return to sender."[55] Reuters News Service quoted delegation leader Robin Hoy: "Another country cannot dump on another that is poorer than them and get away with it. It's a moral obligation we have as a country to take this back." Reuters also reported that "officials at the US embassy in Haiti said they had no plans to respond to the demonstration, calling the argument over the incinerator ash a matter between the Haitian government and a private company."[56] It is curious that the embassy tried to privatize this issue, which raises the question as to whether activists misplaced their focus and might have been more effective had they not focused almost entirely on the obligation of U.S. government officials to act.

Activists continued to push for their goal, often using creative tactics, like delivering samples of the incinerator ash to officials in the United States and Haiti. On Valentine's Day 1998, activists sent two hundred Valentines with pink hearts to Philadelphia Mayor Ed Rendell, which read,

"Have a Heart. This Valentine's Day help bring Philly's waste home from Haiti."[57] Others plastered the area near city hall with signs featuring the mayor's picture and the phrase, "WANTED for Environmental Racism," above his photo.

While it is certain that governments and corporations accused of perpetrating environmental crimes are likely to pay more attention to SMOs from the North, southern SMOs are increasingly playing a major role in global environmental justice efforts. COHPEDA, the principal Haitian environmental organization involved in Project Return to Sender, was increasingly viewed as a legitimate political stakeholder in Haiti. In a letter to Return to Sender members, COHPEDA leader Alex Beauchamps wrote: "Thursday, March 19, there was an 'inter sectorielle' meeting with the Foreign Minister. Present were representatives of several (Haitian) government ministries, the Port Authority, and COHPEDA. The objective of the meeting was to work on the repatriation of the waste. A 'chronogramme' of activities was set, according to which the work of returning the waste would begin in May 1998."[58]

A few days later, Beauchamps reported that COHPEDA was having weekly meetings with the minister of the Haitian Foreign Office to evaluate the state of the cleanup plans.[59] However, concerned that these plans were not moving forward quickly enough, four days later, COHPEDA issued a petition to the Haitian Parliament and the president:

> For the dignity of the Haitian people. For the protection of the environment! Representatives of national and international civil society address the Haitian Parliament to adopt the necessary measures in light of the United States' dumping 4,000 tons of toxic waste on the town of Gonaives. To the Parliamentarians of the Haitian Parliament. . . . This is what we ask the Haitian Parliament: To undertake action before the Executive for the return of the waste back to its country of origin; To study possible legal recourses, given the violation of Article 258 of the Haitian Constitution, forbidding the importation of waste from foreign countries; To ratify the Basel Convention and the Latin-American and Caribbean Regional Accord on the marketing of wastes; Ten years is enough![60]

Thus, while U.S.-based social movement organizations (SMOs) were hesitant to call on the Haitian government to take responsibility for the Philadelphia ash crisis, Haitian SMOs felt this was a necessary approach that Project Return to Sender should use for repatriating the waste. They also refused to allow Haiti to be saddled with the image of an impoverished and passive Third World nation. Haitian SMOs took on both a nationalist and internationalist approach to environmental justice, invoking both the Haitian Constitution and the Basel Convention, and viewed the problem as a human rights violation. As Alex Beauchamps stated, "COHPEDA is an Haitian environmental platform who is fighting for a protected environment, where the *human rights* are respected."[61] The human rights frame extends the traditional "environmental justice frame"[62] to an explicitly international legal framework that is recognized by the United Nations and, in theory, all nations. In an open letter to the people of Philadelphia, Beauchamps elaborated on the contents of the Haitian Constitution and argued that the *Khian Sea* incident reflected international economic inequalities: "The Constitution of Haiti, in Article 258, stipulates clearly: 'No one shall bring into the country any trash or refuse of any kind from abroad.' We see clearly that this dumping was not a question of respect or of not respecting laws, but most of all, this is a question of relations of strength between strong States and weak States, where the dominant interests are those of a fundamentally economic nature."[63]

Beauchamps' letter emphasizes the geopolitical roots of transnational environmental injustices. His organization requested that the Project Return to Sender groups consider how they might elevate the campaign's visibility. This is when Global Response, the U.S.-based transnational network that encourages people around the world to write letters to influential leaders in governments and corporations involved in activities that threaten the human rights and environmental integrity of communities in the global South, entered the picture. Global Response was a member of Project Return to Sender and issued a call for letter writers to target Philadelphia Mayor Ed Rendell.[64] Hundreds, possibly thousands, of letters flowed into the mayor's office within days of Global Response's issuance of the alert. Rendell's staff mobilized to write a form letter response, which was mailed to every person who sent in a letter. Here are excerpts from that letter:

> As to the question of the ash in Haiti, despite the fact that the City of Philadelphia has no legal obligation to do so, the City has offered on numerous occasions to assist those seeking a reasonable solution to this situation. The material left in Haiti in 1988 was not left there with the permission or knowledge of the City or any of its representatives. A subcontractor of the City's contractor transported the ash to

Haiti without the City's knowledge. Upon discovery of the situation, the City withheld payment to its contractor pending proof that the subcontractor had disposed of the ash in a legal and appropriate manner as required by its contract. When that proof was not produced, the City encouraged, supported and provided evidence to the U.S. Department of Justice in their indictment and successful prosecution of the operators of the ship responsible for dumping the incinerator ash in Haiti.[65]

Thus, the city's view of the situation was that it was simply a legal issue involving a breach of contract on the part of the company that shipped the waste. Such a perspective seems naive and shortsighted at best, considering that the ash was Philadelphia's waste, Philadelphia was a party to the contracts, and the public outcry was so great that the mayor's lost political capital likely far outweighed the financial cost of paying for the cleanup.

In keeping with creative and cost-effective methods of protest, Global Response urged letter writers who received responses from Mayor Rendell to "return it [the letter] unopened, writing in large letters 'RETURN TO SENDER.' His office will be inundated with a second round of letters, with the campaign message loud and clear."[66] Many activists acted on this request and returned the letters. This tactic allowed Project Return to Sender members and allies to make use of the U.S. postal service by sending the letter free of charge (constituting a second, albeit smaller, letter writing campaign), and provided the opportunity to remind the Mayor's office of the campaign's slogan and goals.

Local and Global Action around Environmental Injustice

Two other related events were ongoing at that moment: one was a struggle occurring in Philadelphia's backyard and the other an international meeting of representatives and activists from nations around the world to determine the future of the global trade in hazardous wastes.

Just southwest of Philadelphia is the small town of Chester. A local environmental justice organization, Chester Residents Concerned for Quality Living (CRCQL), charged the city administration with environmental racism. They vowed to fight Mayor Rendell's plan to burn 12 percent of the city's trash in their town, which was mostly African American (65 percent) and low income. Philadelphia had stopped burning its own waste in 1987 and was sending it to nearby towns. The nation's fourth largest incinerator, the largest medical waste autoclave, and a sewage treatment facility and sludge incinerator are all located in Chester, just yards from residents' homes. Mayor Rendell had recently declared his intention to sign a multi-year contract, worth nearly $300 million, to send 100,000 tons of municipal solid waste each year to an incinerator in Chester. Zulene Mayfield, president of CRCQL, testified at a city finance committee hearing, "We believe the trash is coming to Chester because we're Black. This is environmental racism." Mayfield noted that Philadelphia has a history of racist aggression, including the police department's infamous decision to bomb the radical MOVE organization's headquarters in 1985 and the city's role in sending incinerator ash by boat to Haiti. Moreover, the U.S. EPA had concluded that children in Chester had blood lead levels that were unacceptably high—and higher than any other community in Pennsylvania. The community's mortality and lung cancer rates were 60 percent higher than the county's averages. Trucks carrying infectious waste and garbage entered the community every four-and-one-half minutes, carrying waste containing carcinogens and other toxics along streets where schoolchildren in Chester walked and played.[67] Hence, the situation in Gonaives, Haiti, was not so foreign to the political and business establishment of Philadelphia after all. They had been smothering Chester with several different toxic waste streams for years.

That same year at the Conference of Parties to the Basel Convention, representatives of the world's nations and numerous stakeholder groups took steps to fully implement the Basel Ban, to disallow waste trading between OECD and non-OECD nations. As Greenpeace's Kevin Stairs put it, "We need to conclude the process we started years ago to ban *waste colonialism* and move into a new era of *clean production.*"[68] Greenpeace and other environmental justice activists saw the waste trade as a clear example of inequalities between nations, and proposed that clean production be a positive step forward.[69] Thus, in the light of the Basel Ban, the U.S. government, which refused to ratify the ban, was increasingly viewed as a maverick and rogue state, particularly while less well-endowed nations attempted to follow the spirit and letter of this agreement. The Philadelphia ash scandal was becoming an embarrassment.

Activists continued to push for a change. The next year, Project Return to Sender members gathered in Haiti for an unprecedented conference and series of strategy sessions around the Philadelphia ash situation in particular and globalization and environmental justice in general. More than seventy activists from sixty organizations based in Haiti, the United States, and Puerto Rico spent three days together in June 1999. They passed a resolution calling on the United States to repatriate the waste and for the government of Haiti to ratify the Basel Convention. They had meetings with government officials to that effect as well. More interesting was what happened outside the conference proceedings. U.S. activists met with representatives of USAID in another effort to convince that agency to support Project Return to Sender. As Essential Action's Ann Leonard wrote to the Return to Sender members afterward, USAID staffers "rolled their eyes when I asked them about the waste and reaffirmed that they are not going to contribute any money towards a solution. One USAID official confirmed that they received instructions from Washington not to get involved in the case because if USAID helped clean up the ash in Gonaives, it would set a precedent that USAID may help clean up the piles of U.S.-generated waste which was dumped in so many countries, including Bangladesh, India, South Africa, and elsewhere."[70] USAID thus made it clear that the Philadelphia ash scandal was a test case for whether environmental justice activists could force the United States to admit responsibility for its global environmental injustices.

Meanwhile, the Haitian public was not complacent about the scandal. At one public gathering, tens of thousands of Haitians in Gonaives protested the delayed return of the ash back to the United States. "It's the biggest demonstration we've ever had" the mayor of Gonaives stated. Protesters carried placards, dressed in white, and shouldered symbolic coffins while chanting, "Gonaives is not a garbage can!"[71]

The Journey Home

After more than a decade since its departure from Philadelphia, the toxic ash was finally removed from Haiti in April 2000, to be returned to an undetermined location in the United States. This was a cooperative effort among the Haitian Office of the President, the Haitian Ministry of the Environment, the U.S. Department of Agriculture, and the New York City Trade Waste Commission. Arpin, a U.S. company, oversaw the loading of the ash onto a ship that left Haiti on April 5, 2000. It is unclear whether Philadelphia contributed anything to the effort. The Haitian government issued a press release on the occasion, giving credit to Project Return to

Sender members for making this event possible, and stating, "This victory of environmental justice is a victory for all poor nations struggling to protect their countries from becoming dumping grounds."[72]

What happened for the next two years was not surprising, but outraged many observers nonetheless. After so casually dumping the waste in Haiti, the United States still had great difficulty accepting that same waste—its own waste—within its borders. Several states rejected the waste, including Florida, Georgia, Ohio, and South Carolina. Not surprisingly, one of the target locations in the United States was a poor community of color: the Cherokee Nation of Oklahoma.

A task force of concerned Cherokee Nation citizens, community residents, and tribal representatives from the United Keetoowah Band of Cherokee Indians formed to oppose a deal that was being developed between the leadership of the Cherokee Nation of Oklahoma and the operators of the barge carrying the waste from Gonaives. A committee of residents living near the landfill was organized and called itself "Don't Waste Indian Lands."

The alleged efforts of Cherokee leaders to bring the waste within its jurisdiction are not surprising, given the popular approach to waste management as a form of economic development, particularly among Native and other communities of color.[73] The Cherokee nation ultimately dropped the proposal, and in July 2002, the waste was finally deposited in the Mountain View Reclamation Landfill, about 120 miles outside Philadelphia. This was the end of a sixteen-year journey that involved attempts to deliver the ash to nearly a dozen nations' ports, being turned away from some of those ports at gunpoint, a near mutiny by the *Khian Sea* crew and engineer (the latter of whom was jailed in Yugoslavia after threatening to scuttle the ship), and being refused by six U.S. states and the Cherokee Nation.[74] The landfill is in south-central Pennsylvania in Franklin County, near the town of Greencastle, which is a mostly Anglo American city (96 percent) with moderate income levels ($41,000 compared to the national average of $42,000).[75] Ultimately strong environmental justice battles sometimes result in the waste being shifted back to white, middle-class communities. Perhaps these Return to Sender operations do constitute a form of environmental justice, and perhaps when more white or otherwise privileged communities have to shoulder these toxic burdens, the nation might become more attentive to the issue.

Greenpeace and Corpwatch activist Kenny Bruno remarked on the environmental injustice involved in the ash case: "If, for some reason, incinerator ash had been dumped on a beach in France or Britain, Philadelphia would have found a way to get rid of it within a few hours. But because it was Haiti, it sat there for 12 years."[76] A Lebanese émigré named Winnie lives in the town of Seguin, Haiti. He has created an environmentally protected zone in this city and is well aware of what happened in Gonaives. He declared:

> America cannot be the leader of the world—claim it, boast about it—and let things like that happen. It's not moral; it's immoral. . . . Why just go and dump it in the poor guy's backyard? Because it is a poor guy's backyard and he cannot do anything about it. How many Haitians will die because of that? . . . Is Haiti another planet? People should come and visit this planet.[77]

These statements by Bruno and Winnie speak to the reality and perceptions that come from segregation, separation, and "othering" that go beyond spatial and geographic differentiation, to reflect early-twentieth-century sociologist Robert Park's concept of social distance,[78] because although Haiti is a neighbor of the United States, it might as well be, as Winnie put it, "another planet."

Back to Gonaives

Although the ash was removed from Haiti, its effects were long lasting, as is often the case with the hazards associated with a risk society. Raphael Elifaite, a resident of Gonaives, was one of the workers hired to help remove the waste in April 2000. He and other workers reported various physical ailments in the wake of the removal: "After this work, I wake up in the middle of the night. My nerves feel bad. And the rash, it is bad. I really scratch and scratch and scratch. And also spinning, dizzy. There is something with my nervous system. Something attacks my nervous system. I feel depression I never felt before this."[79]

Elifaite's symptoms are strikingly similar to those suffered by the workers who helped unload the waste from the *Khian Sea* when it first arrived in Haiti in 1987. Kend Noel was the first Haitian known to have contact with the ash. He later stated, "We all have a rash. We do not feel comfortable. We are not normal. We feel that something happened. My head is not good. It was windy at the wharf. I was breathing this thing. It affected me."[80]

As environmental justice activists and some scholars have often noted, community-level toxics affect not only ecosystems and residential areas but also workers, who are often the first, last, and most directly exposed to toxics. Thus, environmental issues are labor issues as well.[81]

Some observers have commented that the international trade in hazardous wastes may not be so significant a problem as activists make it out to be because the United States exports only 3 percent of all the hazardous waste produced domestically, and most OECD nations traded hazardous waste with each other even before the Basel Convention.[82] However, according to Montague:

> Today the U.S. maintains no records of most exports of toxic waste because most of it is exported in the name of recycling. Once a waste is designated as "recyclable" it is exempt from U.S. toxic waste law and can be bought and sold as if it were ice cream. Slags, sludges, and even dusts captured on pollution control filters are being bagged up and shipped abroad. These wastes may contain significant quantities of valuable metals, such as zinc, but they also can and do contain significant quantities of toxic by-products such as cadmium, lead, and dioxins. Still, the "recycling" loophole in U.S. toxic waste law is big enough to float a barge through, and many barges are floating through it, uncounted. The prevailing attitude seems to be, the U.S. has a right to dump on the rest of the world.[83]

Thus, the 3 percent figure may be misleading and bear little resemblance to reality. The problem of toxic exports remains a serious global issue because it continues underneath the regulatory radar, just as it did in the early days. While the "sham recycling" Montague refers to is a major problem with municipal solid waste, this practice has also been used where other waste streams are concerned, such as electronic waste (see chapter 6).[84]

Many of the leaders of the Project Return to Sender coalition were confident that the solution to the international waste trade was recycling and clean production. Consider the following statement by the coalition: "As long as cities can export their waste and never have to take responsibility for it, they will never have incentive to deal with their waste in a sustainable manner—through waste reduction and recycling." This proposal is put forward despite the fact that environmental inequality is the result of the normal functions of political-economic and cultural systems, while recycling and clean production are largely technical approaches to a resource management problem. In other words, the long-term solutions to the problem may not entirely match activists' short-term political responses (which are in fact appropriate and well reasoned). That is, in their initial campaigns to repatriate such wastes, activists underscore that the waste trade is the result of the exploitation of both people and the environment, yet in their vision of a better future, the role of social and political equality seems to fade into the background in favor of a primarily engineering perspective. Clean production is not going to solve the problem of environmental injustice if social inequalities are the root of that problem.

Nevertheless, the Philadelphia ash case represents an important milestone in the development of the global environmental justice movement. It was a testing ground for Greenpeace in its efforts to develop collaborative relationships with national governments and nongovernmental organizations in the South. As Kenny Bruno later stated, "Greenpeace . . . had offices in a lot of countries . . . and we also had contacts in other countries. Like . . . in the case of Haiti, we were contacted by an environmental group down there called Friends of Nature, and later the lead group from Haiti changed to a group called COHPEDA, and we simply collaborated with them on a collegial basis. We shared information, we shared strategies, and we worked together. They told us what the government down there was saying, we told them what the government here was saying, and so on."[85]

While most observers agree that the *Khian Sea* case provided the impetus necessary to develop the Basel Convention on the Control of Transboundary Movements of Hazardous Wastes, Kenny Bruno made it clear that the case also led to the development of Greenpeace's toxic waste trade campaign:

> We found, when this case came to light, that there had been, during that same period, a famous case of a garbage barge that wandered the seas for about three or four months in the summer of 1987. And at the same time, our Greenpeace colleagues in Italy discovered a case where PCBs from Italy were being stored in Nigeria. And "stored" is a kind word. I mean they were out open in the sun leaking in the barrels. And we just started looking into it and finding a pattern that this was happening more and more and that there were dozens, even hundreds of proposals to send the ash and other materials legally. . . . There were dozens, even hundreds, of other schemes and so it certainly woke us up and we kept looking into it, and as our name became associated with the case, people would send us documents and send us plans of other schemes like this, and . . . we discovered that there were hundreds of them. So we organized a campaign to uncover them and to oppose the export of toxic waste.[86]

Greenpeace's global toxic trade campaign became a blueprint for other international environmental justice actions in the years that followed. Sophisticated protest campaigns by global environmental justice networks have since persuaded Italy to retrieve toxic waste from Lebanon and Nigeria; Germany from Albania and Romania; South Korea from China; the United States from Bangladesh and India; and Japan from the Philippines, to name only a few other examples.[87] We can now add "United States from Haiti" to that list. There is a sign on the side of the road leading to Gonaives, Haiti, that reads: "Toxic Dumps in Haiti: Never Again."

Ban the Burn: The Anti-Incinerator Movement in the Philippines

In June 1999, the Philippine government did what no other nation had done before: it banned all waste incinerators in the country. The Philippine Congress approved the Clean Air Act, which allowed a total ban on waste incinerators. Social movement organizations around the world hailed the decision as a major triumph for environmental justice, particularly since this progressive legislation emerged from a global South nation.[88] Filipino activist and Greenpeace toxics campaigner Von Hernandez declared, "This decision sends out a very strong signal against the dumping of toxic technologies on developing countries like the Philippines."[89] Hernandez stated that the law was "an inspired move, which recognizes that development in the new millennium must be founded on sustainable and non-polluting processes and industries. The real solution to Asia's waste problems is to jumpstart waste reduction, recycling and clean production programs."[90] The decision to ban incinerators was also in line with the growing movement that seeks to phase out persistent organic pollutants (POPs) such as dioxins and furans, which are by-products of incineration.

The Philippines Clean Air Act came at a time when the multibillion-dollar incineration industry was (and still is) confronting shrinking markets in the more industrialized nations, prompting various incinerator companies to "enter greener pastures in the developing world."[91] During the early 1990s, a French company, Générales des Eaux, planned to locate what would be the world's largest municipal waste incinerator at San Mateo town, north of Manila. However, as a result of protests from activists and local citizens, the plan was rejected. Activists were keenly aware of the political economic opportunity structure and how it revealed collaborations between state and industry actors who were often pitted against communities and environmental justice advocates. As Odette Alcantara, an activist with the Filipino environmental group Mother Earth Unlimited, noted, "Unless we keep a strict vigil, the incinerator lobby may force the government to change its mind on the ban in a moment."[92]

Anti-incinerator activists and the incinerator industry have diametrically opposed views on the subject of the ecological impacts of waste burners. Von Hernandez makes a standard environmentalist-scientific claim about the fact that most incinerators produce dioxin and that "there is no safe limit, no 'general standard' for dioxin exposure. The chemical is known to be about 200,000 times more toxic than DDT."[93] Incinerator company officials, for their part, firmly insist that their machines are safer and cleaner than at any other time in history. William Schaare of Industron Inc., a firm that manufactures incinerators, claims that "state-of-the-art incinerators now are virtually pollution free. The dioxins emitted by 1,000 incinerators today are equivalent to those emitted by one incinerator in the 1980s."[94] Both may be right, but if there is no safe level of dioxin exposure, as many scientists conclude, then the activists may have the stronger position.

The lesson from the Philippines' Clean Air Act is not only that such a piece of progressive policymaking is possible in an impoverished global South nation, but more important, such policies are likely to be proposed and passed only when massive grassroots community mobilization and social movement action are evident. Moreover, the success of social movement organizations is that much more significant here because they remain "continuously under attack by both industry and international agencies such as the Asian Development Bank, which see it [the Clean Air Act] as an affront to their commercial interests."[95] This was largely because, as news spread of recent garbage management problems in Manila and other major urban centers in the country, "multinational waste management companies targeted the Philippines because they saw enormous business opportunities."[96] These companies sometimes had assistance and support from foreign diplomats such as the Swedish and Danish embassies, powerful financial interests like the U.S. and European Chambers of Commerce, and banks and multilateral aid agencies like the Asian Development

Bank and the Japan International Cooperation Agency. These institutions joined forces with incinerator companies in an attempt to block the incinerator ban. In the run-up to the vote, they took lawmakers on expensive trips to European countries to show them how "clean and safe" incinerator technology can be.[97] They sent word to the Philippine government cautioning against endorsing the proposed legislation because of possible future World Trade Organization sanctions.[98] Thus, as activist Von Hernandez stated, "All establishment forces were against it."[99] Not to be deterred, Filipino environmental groups created the Clean Air Coalition, which presented Congress with more than 2 million signatures supporting the incinerator ban.

Von Hernandez remembers that Greenpeace was one of the first large SMOs focusing on this issue in the Philippines. Greenpeace had an ongoing global campaign against incineration, so this case fit well into the organization's larger agenda: "We had the materials and we knew what was happening in Europe, and we knew about opposition in the United States to incineration. And we discovered that there were about ten proposals to erect incinerators in metro Manila—including American companies like Ogden Martin, German companies, and Japanese companies. There was also a Belgian proposal to build an incinerator for free, as long as they would accept waste from Belgium. These proposals were coming up and our government officials were taking them in a positive way."[100]

Thus, even if the waste in question is generated domestically, the fact that TNCs and foreign firms seek to sell their products to southern nations is an environmental justice concern because incineration is a hazardous technology. So whether through dumping waste or technology transfer, the global South faces ecologically perilous pressures from the North.

As is often the case in politics, when special interests offer financial incentives to interested elected officials, the latter can frequently find a way to push an agenda forward. Generally the Philippines is no exception. As Hernandez recalled, "The presidential election was coming up, and we exposed the corruption involved between this incinerator proposal and huge kickbacks to officials, direct bidding that happened, and approving the construction of this incinerator. We exposed it and it became an election issue. The dominant opposition party took it up and adopted an anti-incineration stance. The opposition party was then headed by the popular movie actor Joseph Estrada, and he won by a huge margin. When he won, again we campaigned for approval of the Clean Air Act with the incineration ban provision intact, and he finally signed it."[101]

Activists also drew on more innovative social movement tactics, focusing on community-based organizing across multiple areas, so as to plug up the holes in which incinerator proponents might find a place to locate their burners. This was the movement's response to the "not in my backyard" (NIMBY) approach that characterizes some community responses to incinerators. If one community rejects an incinerator, the developers may try to sell it to a nearby community. This is a losing game for residents, who see only their local interests at risk. As Hernandez's colleague Ann Leonard likes to put it with regard to the global anti-incinerator struggle, "We consciously challenged the NIMBY model of politics and moved toward the NOPE—or the Not On Planet Earth—approach."[102] Von Hernandez explains this in the context of the Philippines struggle: "What we did was, whenever these proposals came up, we hooked up with the community that was being eyed to host the facility, and we organized these communities together to form a united front against the incinerator proposals. That was the first step. But we realized that we had to be much bigger than just a united front of communities opposing incineration, we had to involve other sectors like the Catholic Church, which is a powerful political player in the Philippines, a largely Catholic country. We also linked up with women's groups, with other health and public interest organizations, and we did lobbying."[103]

With regard to lobbying, Hernandez also brought a unique advantage to the movement effort: "Because I worked in the Philippines Senate before as a legislative chief or senator, I knew how the process worked in legislation. So we used our contacts in the Senate and also in the House of Representatives. We have a similar model like you have in the U.S., and we lobbied both houses. Just public pressure and pressure politics, and at one point we got the incineration ban provision in the legislation and there was a big fight in congress. But we simply defeated the arguments put forward by the proponents, based on science, on economics, and on the environmental issues. They just couldn't beat us on those issues."[104]

Activists also worked at securing change on other levels and scales. A popular saying among activists is, "The price of success is eternal vigilance."

Activists in the Philippine Clean Air Coalition know this all too well, because the instant they succeeded in banning incineration, industry and government forces stepped up to undermine the law and actively promote waste burners. Greenpeace, GAIA (which formed in 2000 in the Philippines and lent support to this struggle), and the Clean Air Coalition also knew that the best way to achieve a lasting ban was to strengthen community organizing at the grassroots as well as force the state to acknowledge that the law was in compliance with the Philippine Constitution and the UN Stockholm Convention. They drew on policies, conventions, laws, and regulations at several different scales to make the argument.[105] This tactic was productive, as Von Hernandez noted:

> It is very effective because it exposes the contradictions in government policies and programs and, for example, international law conventions that we have agreed to as a country. For example, the Basel Convention, the Stockholm Convention, and even the Earth Summit Articles—Agenda 21—and all that. As far as we're concerned, when governments of the world adopted these conventions, they have become the global standard, what we call the "least common denominator." I mean we could go much further than the standard, but right now this is what all governments have agreed to. In the case of the Stockholm Convention, in fact the aim to ultimately eliminate dioxins and furans we think is a very crucial objective because it could ignite the clean production revolution that we have wanted to happen globally. But that of course will depend on how things get implemented on the ground. They are the standard and governments have agreed to adopt them and therefore they have an obligation.[106]

TSMOs deftly use multilateral and international environmental agreements to further their global aims of environmental justice and to push particular states to phase out incineration. What is also clear is that these activists consciously take part in shaping these agreements from the beginning (where possible) and then take a serious role in implementing and enforcing them. Jorge Emmanuel, a Filipino scientist and medical waste incineration expert who consults with both TSMOs and the World Health Organization, is deeply involved in efforts to promote safe alternatives to incineration. On the topic of activist impact on international agreements, he stated:

> There are international conventions that the U.S. is a signatory to that we hang on to and say "you guys signed this and therefore you have a responsibility for what is happening." The Stockholm Convention on POPs is an example. This was a convention where the NGOs had a big role in its development . . . and we use it a lot

because the Persistent Organic Pollutants include, for example, dioxin and, of course, that means incineration because incineration is the major source of these dioxins in the global environment. And then, of course, it lists what they call "the dirty dozen"—all these other pesticides and other nasty chemicals that should be stopped. Under the Stockholm Convention, countries have four years to come up with an action plan to phase out these systems. There is still a lot of grace period. So that's why the NGOs are specifically involved and we are using them to work with governments to be able to push them, because we're saying why invest in an incinerator when, under the Stockholm Convention you'll have to phase it out anyway? So that is the position we're taking. Because I have to convince the people that you shouldn't go back to incinerators and then I use the Stockholm Convention as one of the reasons to do this.[107]

Eternally vigilant, the Clean Air Coalition in the Philippines pushed further, knowing that opponents would pursue other means to achieve their goals. Von Hernandez recalled:

After we got the incineration ban, we did not stop there. We knew the next threat would be landfills. And we had a lot of landfills coming up. Our opposition to landfills is not really fundamental, but only in the sense that it would impede the mainstreaming of what we felt was the lasting solution to the problem: recycling, composting and source separation and, of course, on top of that, waste reduction. But without mainstreaming that first, we would be facing a deluge of landfills, mega landfill proposals.[108] What we did was a similar strategy with the anti-incineration campaign. We linked up the anti-incineration groups with the anti-landfill groups, which were mostly working under a network of the Catholic Church, and expanded our coalition so that it later became the Eco-Waste Coalition in the Philippines (it used to be the Clean Air Coalition). We again lobbied for the passage of an Ecological Waste Management Act, which would prioritize recycling, composting and source separation, and set recycling targets and also some packaging directives, and we got that a year after we got the Clean Air Act. So they haven't stopped, they bring up this option of incineration saying, "We have a disposal problem, we need to amend the Clean Air Act." And there are such bills pending in Congress today calling for a repeal of the Incinerations Act.[109]

The Clean Air Coalition broadened its goals, and its framing of the problem and solution, and expanded its membership, transforming itself into the Eco Waste Coalition, a national alliance of more than one hundred organizations campaigning for the implementation of ecological waste management programs, which sought to implement the kinds of practices that would support the incineration ban. Without the institutionalization of recycling, composting, and waste separation, there would be little hope for the ban's success. With laws, of course, there is the initial act of passage and signing, but then implementation is required. Here is yet another dimension where the importance of a strong local movement presence is clear, because it is often critical to enforcing and implementing laws.[110] Consider Greenpeace's actions in this regard: "We did neutralize the mayors in 2001, when they really came out in force calling for this revocation of the ban. What we did—this is Greenpeace—we took garbage from their homes and also from their offices and then exposed that to the media just to stress the point that these mayors were the first to violate the law. We have an Ecological Waste Management Act, which mandates recycling and source separation and you have the mayors in their own households not doing separation. What hypocrisy! They did shut up after that [laughs]. They were furious, of course, but the public loved it."[111]

In that guerrilla theater action, Greenpeace and allied activists from the Eco-Waste Coalition in the Philippines accused the Metropolitan Manila Development Authority (MMDA) and a number of Metropolitan Manila mayors of hypocrisy with regard to addressing the garbage crisis, particularly since the laws passed require responsible consumer and institutional behavior. In a dramatic act of grassroots theater, activists presented the unsegregated (meaning mixed, and not ready for recycling) household waste of several key metropolitan officials before newspaper and television cameras. In a press conference, the group revealed the rubbish taken from the residence of Pasay City Mayor Peewee Trinidad, the neighborhood of MMDA chair Benjamin Abalos, and the city hall of Manila Mayor Lito Atienza in order to call attention to their violations of (and, by extension, the general problem of government officials ignoring) the Ecological Solid Waste Management Act of 2000. Von Hernandez was on the scene and declared, "No wonder we have a garbage crisis. The officials who are supposed to implement the law are guilty of violating it themselves."[112]

This is one of the hallmark tactical approaches Greenpeace is known for: actions that are creative, symbolic, nonviolent, and with a flair for the dramatic, often using street or guerrilla theater that frequently attracts media attention. Overall, however, the Eco-Waste Coalition divided up the tasks for different member organizations based on what strengths each group brought to the struggle. And although Greenpeace does an excellent job of staging high-profile events, its main role was actually technical support.

Francis dela Cruz, the leading Greenpeace campaigner for the Southeast Asia region, put it this way:

> The role that Greenpeace played was more the scientific backup for the policy proposals that were put forward. I think Greenpeace was in a very good position to marshal resources—both technical and scientific—from its defense offices, and I think that helped the debate. Of course politicians in the Philippines I suppose are just like politicians in other places: they have to see numbers, and that is where local groups played the major role. It was a multi-sectoral effort. We had people coming from different economic backgrounds. I remember a group of largely upper-middle class elderly women who would just go into Congress and Congressmen just listened to them. But it played out really well because we had people who could go in and work inside the halls of Congress and we had people who exerted pressure from the outside. Greenpeace doesn't have that many people; we have volunteers. But we cannot mobilize thousands of people, so Greenpeace provided a lot of the technical stuff. So it was really a joint effort. And I suppose the fun part of it was being able to use to our advantage these different strengths. Local groups had the numbers, Greenpeace had some technical staff up here, so just employing them at the right time and place—that was the story of the incineration ban.[113]

The presence of Greenpeace aided Filipino activists in elevating the visibility of the local and national struggle for environmental justice there. It also provided protection against the kinds of harsh state repression that many activists there had come to expect, since their days as student movement organizers against the government of Ferdinand Marcos. Dela Cruz recalls:

> Here, if you're an activist organization, you are almost expected to get your head hit. Greenpeace has done actions here and we've never been hit by the police. We've been arrested a few times, but even the police respect us. I distinctly remember organizing and negotiating, being the police liaison for an action at the U.S. Embassy. And to us, if you go to the embassy, you'll most likely get your head bashed in because they're so used to all these demonstrations turning violent. But even the police officer told these men, "Oh, this is a different lot, this is Greenpeace, they're not the same as the other guys, so be gentle." And they were very gentle. Everyone got arrested, but nobody got hurt. Even the police were quite happy—I think—to see us in action. They know that we'll do something, we'll get our message across, and nobody will get hurt. I think that is our positive contribution to the protest movement here. You can protest, and do what you want, but not end up in violence.[114]

Among other roles, TSMOs can offer an extralocal protective shield if local authorities believe there will be an international political price to pay for repressing members of such groups. Greenpeace and other

TSMOs knowingly play this part in order to advance their agendas and encourage domestic authorities to adopt what they view as progressive policies.

Medical Waste Incineration in the Philippines

One of the most visible and infamous embodiments of the state-corporate nexus that produces environmental inequalities globally is the World Bank. Governments fund the bank, but it operates in conjunction with large corporations to support development projects around the world. Opening these nations up to free market principles is the objective, which often results in exposing them to greater exploitation by foreign corporations. Funding the incineration of municipal, medical, and hazardous waste in the global South has been one of the World Bank's favored approaches in recent years.

Prior to the incineration ban, many Filipinos had faced the threat of incineration from before they were born and throughout the rest of their lives. The human fetus is exposed to dioxin and other POPs produced by medical waste incineration, a routine component of many hospitals in the Philippines. Jorge Emmanuel related a story about the impact of state inertia in the implementation of the incinerator ban and how this affected some of society's most vulnerable members:

> In the late 1990s, the Austrian government gave the Philippines a loan to purchase twenty-three state-of-the-art incinerators from Austria. I returned to my home province around that time and saw one of these incinerators. I walked in expecting a real state-of-the-art incinerator and, my god, it didn't have any air pollution control at all and it's one of these old ones, like a 1950s or 1960s era design. But what was emotional for me was when I went to my family's province and saw the incinerator, it was like thirty feet away from the hospital. And of course the hospital—not like our hospitals in the U.S. that have air conditioning units—they just opened the windows for air conditioning. I said, "Well, it is so close that all of that toxic stuff is going to go into that window right there and if there are any patients there then they're going to be affected." So we went up there and guess what that place was? It was the neonatal ward, the most sensitive of all patients, premature babies, and they're breathing all this ash that's coming in from the incinerator. So the environmental groups actually videotaped it, where you can see the ash from the incinerator coming into the neonatal ward.[115]

Newborn children were breathing toxic incinerator ash in their first hours of life, and the source was ash from a second-class incinerator that a global North government would not allow in its own borders but would willingly sell to the Philippines. With the incinerator ban in effect, activists will try to push the state and the citizenry even further from the incinerator industry, which has held sway over the nation's waste management systems for so long and has exceptional influence in Southeast Asia.

External Pressure, Eternal Vigilance

Francis dela Cruz is a toxics campaigner for Greenpeace, based in Southeast Asia. He is part of a regional office in Bangkok, Thailand, and spends half of his time there and half of his time in the Philippines; he focuses on waste incinerator proposals in many countries. There is a major push by incinerator developers to move into the region, since rich nations like Japan are unable to build burners at home owing to the heavy politicization of the issue there. Japan is infamous for incinerating much of its waste in the past, and as environmental consciousness has risen in that nation (as has been the case throughout other northern nations), developers have pushed the practice southward into poorer countries. Dela Cruz stated: "Japan was the incinerator capital of the world until now . . . because a lot of incinerators in Japan are being closed down, so they will need markets for their technologies. It is a dying technology in the northern countries, in the global North, so I think that because it's dead up there, then they're looking down here to revive the market. They are always on the lookout."[116]

Regarding the broader causes of international pollution transfers to Southeast Asia, dela Cruz argues that it is the scramble for modernization and discrimination against the poor that produce environmental inequalities:

> I think it can be said that countries like Thailand and the Philippines and countries in the Mekong Delta, all these countries, we come from a region that is very fast growing, and within that region there is a lot of competition among the nation states. And I think in that mad scramble, we're just about ready to take in anything that might bring us closer to modernization. The other problem arises where the government is weak or unable to provide for their citizens. The commercial pressures, they just come on very strong. It's true, that for polluting industries, that is just the situation they are looking for. So of course it's discriminatory; they wouldn't put up an incinerator in a middle- or upper-middle-class neighborhood, because they know that these people can articulate themselves, might have a connection here or there. They'd rather put it up in some far-flung province away from the public eye. Of course, even if they know that it is harmful, it comes to a situation where they would say, "It's either I eat today and maybe die later, or I don't eat

today, think about the environment, and die tomorrow." So that is very discriminatory on the part of industry. And it just so happens that our government is not able to exert its mandate.[117]

Thus, according to dela Cruz, strong corporations, weak states, a lack of social capital, and the presence of poverty are driving regional environmental inequalities.

Jorge Emmanuel argues that there are two main drivers behind the spread of incinerators, particularly medical waste burners, in the global South. The first is the assumption that burning waste is "the best way to deal with infection." The second is a combination of profit making and a double standard by which northern nations and companies willingly transfer hazardous technologies to the South while rejecting them in the North. Emmanuel states that there is "a commercial interest in that these are companies that can no longer sell their technologies in the developed countries, so they end up promoting them in developing countries. And to me it is a double standard because you're saying it is bad here, but it is okay over there."[118]

In addition to the Southeast Asian region, industry is pushing incinerators across the global South more generally. Jorge Emmanuel explains:

> The trend is there. . . . The U.S. went from 6,200 incinerators in 1988 and it's now down to 115 for the whole country, and it's even going less and less. And the same in Europe: we documented what was happening in Europe and Canada and then we see the trends in developing countries. . . . I just got an e-mail today: some people in Guyana are asking for help because the World Bank is about to fund a whole bunch of incinerators there. In Africa I was getting all these stories: for example Uganda was about to come in with something like sixty incinerators, and Kenya was going to have a number of them. And when I was in Cambodia for another World Health Organization meeting it was the same: the number of incinerators they were going to put up was just incredible, like two hundred or something. We found out that they are going to bring in hundreds of incinerators into China, so the trend is clearly bad. We're shutting down incinerators in the industrialized world—if that is the right term—the U.S., Canada, and Europe, and it's blossoming in the South and in developing countries.[119]

The Philippines remains at the center of the Southeast Asia/Pacific region's garbage wars, and many observers are looking to that nation as the best indicator of which direction the region will go with regard to waste incineration. Despite considerable pressure from political and economic forces at all geographic scales, the Eco-Waste Coalition was successful at passing the Clean Air Act and banning the future construction of waste incinerators in the Philippines. This is a major victory for local, national, regional, and global environmental justice.

Pepsico in India

In the mid-1990s, transnational activist networks geared up to shine the public spotlight on the tons of plastic bottles finding their way from the United States to India. This was not a typical waste-dumping scheme, but rather a more complex set of processes that nonetheless culminated in a high volume of plastic waste being located in this southern nation, affecting residents, workers, and the ecosystem.[120]

Years earlier, PepsiCo, Coca-Cola, Seven-Up, and other U.S.-based soft drink companies had collaborated to create the Plastics Recycling Corporation of California (PRCC), a firm that facilitates the export of plastic waste from the United States to other nations. PRCC purchases plastic waste from municipal recyclers in the United States and sells it at a reduced price to markets mainly in Asian nations. Pepsico and the other beverage companies subsidize the firm with financial contributions. The result was that in 1993 alone, PepsiCo exported more than 9 million pounds of plastic waste from California to Madras, India. Raising concerns about this trend, Indian SMO leaders teamed up with Greenpeace activists to investigate a plastics recycling operation outside Madras that was receiving a great deal of this waste. Futura Industries was the site of significant plastic bottle recycling efforts, yet much of the plastic was simply being dumped on site:

> "As we came over the hill in our auto-rickshaw, we saw a mountain of plastic waste," recounts Madras environmentalist Satish Vangal, one of the researchers who discovered the site. "Piles and piles of used soda bottles stacked behind a wall. When we got closer to the factory, we found many bottles and plastic scrap along the road and blowing in the wind. Every bottle we saw said 'California Redemption Value.' They were all from California's recycling program and now they are sitting in a pile in India!" explains Vangal. "We have enough problems dealing with our own plastic wastes; why should we import other people's rubbish?". . . Vangal explains that "instead of a Pepsi bottle going to a dump in California, a few months later you have a piece of polyester going to a dump here in India. This isn't recycling. At best, this type of reprocessing delays the eventual dumping of the plastic. At worst, it encourages consumers in California to buy more plastic, since their environmental concerns are lessened by the promise the bottles are being recycled."[121]

Indian activist Satish Vangal's account speaks to the concept of delayed dumping, which we also see in the discussion of electronic waste dumping in chapter 6. What is insidious about this type of global environmental inequality is that it can easily be framed as a form of economic development when in fact it is a way of passing on commodities in their last useful stage to a poor nation, where low-income persons will have to manage these products and soon after dump them in their nation.

Many Madras citizens declared their opposition to this practice, and Pepsi became the main target. Hundreds of activists defaced Pepsi billboards in India, calling on Pepsi to "quit India" and calling on India to "*Pepsi Hatao*" ("Evict Pepsi").[122] More specifically, regarding the environmental injustice dimensions of this case, Vangal stated, "Westernized countries should worry about reducing their consumption of plastic, instead of quietly moving their hazardous plastic factories and shoveling their plastic wastes into my country."[123] Greenpeace activists concurred and denounced the project as an example of a U.S.-based corporation avoiding strict domestic labor and environmental regulation by locating toxic operations in poor nations. Marcelo Furtado, a toxics expert with Greenpeace in Brazil, pointed out that "single use plastic soda bottle production is an example of a hazardous and totally unnecessary technology. Instead of shifting this polluting technology to the Third World, Pepsi should bring back clean, safe, refillable glass bottles, like those used throughout India."[124]

Soon after this case made the news in India, a transnational coalition of activist groups joined together, including Public Interest Research Group (PIRG) in Delhi,[125] Samajwadi Abhiyan (an Indian NGO), Greenpeace, and Global Response. At the request of Indian SMOs, in July 1994, Global Response issued an action alert to its membership in dozens of countries, urging them to write letters of protest to PepsiCo's chief executive officer, Wayne Calloway. Global Response collaborated with Greenpeace and Samajwadi Abhiyan on this alert, encouraging letter writers to "ask Mr. Calloway to respond to charges by Indian environmentalists that PepsiCo is exporting polluting technologies from the United States to India" and to "urge PepsiCo to take a leadership role in returning to the use of safe; non-toxic, refillable glass bottles." The action alert also stated:

> Global Response letters are requested to protest PepsiCo's plans to expand its toxic plastic waste discharges into India's environment. The plastic waste that is recycled

is processed in a manner that is hazardous to both Futura's workers and the environment. PepsiCo has now received permission to build a plastic bottle manufacturing plant near Madras. Indian environmentalists consider this plan to be further proof that PepsiCo is turning India into a "full-cycle" dumping ground for PepsiCo's toxic plastic waste. Disposable plastic bottles, with their hazardous manufacturing byproducts, are to be produced in India. The bottles will be shipped to Europe and the United States. Used bottles will return to India as plastic waste. The plastic waste that is not dumped or burned as garbage will be reprocessed into polyester, polluting the environment and the workers at the reprocessing plant.[126]

I note here the spatial link among the pollution of ecosystems, communities, and workplaces that these activists articulated and protested, thus bridging the work-environment divide that many environmental organizations fail to engage.[127] This is critical since so few environmental groups in the global North view labor and the work environment as part of their struggle. The action alert then called attention to the volume of waste being shipped from the United States to India and underscored that this was a case of environmental injustice:

Greenpeace International . . . found that Pepsi exported [from the United States to India] about 4,500 tons of "plastic scrap" bottles in 23 shipments in 1993. According to Futura representatives, the importing agency based in Madras, since 1992, 10,000 metric tons of plastic waste have been imported in India. They estimated that out of this only 60% to 70% can be processed in their plant and the rest 30% to 40% is not reprocessable. Thus, 3000–4000 metric tons of plastic garbage has also been imported and dumped on the Indian soil. The waste exported by Pepsi and imported by Futura is just an example of a much larger trend for western countries to use South Asia as a waste dumping ground. This project will create serious environmental and health problems in India.[128]

When waste is transferred or dumped across borders, workers have to manage, handle, and clean it up, raising questions about occupational health and labor rights. In this instance, "The plastic waste that is recycled in India is processed into polyester under quite hazardous conditions. The majority of women laborers, who are employed to sort and wash the plastic bottles at Futura are paid a very meager sum [30 cents each day]. Futura does not provide them protective clothing or masks to protect them either from scalding water or contaminants, or even exposure to the toxic fumes released during the recycling process. Skin and respiratory ailments have been associated with exposure to the plastic recycling discharges."[129]

Furthermore, gender exploitation was apparent in the factory. Greenpeace activist Ann Leonard reported in a published article that "T. Ganadharam, managing director of Futura Polymers, says he prefers to hire women because 'they are more adept since they have picky fingers.'"[130] This is a claim often made by employers around the globe in a range of industries, from textiles, to canneries, and electronics, but it is frequently an excuse for a much more political reality: typically firms can more easily exploit and pay women much less than men.[131]

Linking the local to the global and elevating struggles in municipal or national spaces to an international scale, the Global Response action alert continued, "India is a party to the United Nations' Basel Convention which prohibits waste trade between parties to the convention and non-parties unless the countries have a separate waste trade agreement in place. Since there is no waste trade agreement between India and U.S., the continuous export of plastic wastes by Pepsi is a violation of international law."[132]

At the same time that Global Response and Greenpeace were coordinating their international letter-writing campaigns, local activists within India coordinated direct action efforts to force the issue:

> Samajwadi Abhiyan [an Indian SMO] has launched a campaign against Pepsi in India. As part of the campaign they court arrest daily outside the parliament. This has entered its 253rd day today. On September 1, 1994 the activists of Samajwadi Abhiyan protested against the defacing of historical monuments. They saw to the immediate removal of the neon signs and logos of soft drinks put up by Pepsi in front of Safdarjung Tomb [an important cultural symbol], obliterating its view and spoiling its architectural aesthetics. Regular defacement of MNC [multinational corporation] soft drink hoardings take place as political boycott action. A number of citizens groups, NGOs and people's movements are active in the campaign.[133]

This action speaks directly to the political economic opportunity structure model, wherein SMOs focus their energies on states and corporations that wield power and drive policymaking. In this case, activists targeted these institutions not only for environmental and human health violations, but also for their impacts on important cultural symbols. This is a classic illustration of a range of problems associated with economic globalization and the depths of people's responses to them.

The backdrop for much of this conflict was the exponential increase in the use of plastic bottles by consumers in the United States during the 1980s and 1990s as the industry made a conscious shift away from using glass bottles toward plastics, which were cheaper, lighter, and easier to transport. Greenpeace reported that in 1972, U.S. industry made 288 million plastic soft drink containers.[134] In 1987, there were 6.5 billion, an increase of more than 2,000 percent. By the beginning of the 1990s, more than 7 billion such bottles were manufactured in the United States alone.[135] This was an unprecedented increase.

Plastic is much more harmful to the environment than many consumers may believe. There are several toxic by-products associated with plastics production, which include ethylene oxide, benzene, and xylenes. These chemicals can contribute to and cause cancer and birth defects and damage the nervous system, blood, kidneys, and immune systems. Global Response and other NGOs in the coalition urged PepsiCo to consider shifting back to the use of glass bottles because "unlike the chemicals used in plastic production, these materials [used to make glass] are solid, inert, non-flammable and largely non-toxic. The production of a 16–ounce glass bottle results in 100 times less pollution than is produced by one 16 ounce plastic polyethylene terephthalate (PET) container."[136]

The activities of NGOs like PIRG and Samajwadi Abhiyan, combined with pressure from Greenpeace and Global Response, seemed to yield results. As Kavaljit Singh (of PIRG, New Delhi) wrote to Malcolm Campbell (executive director of Global Response), "Our timely intervention forced the Pepsi chief not to announce the future plans of Pepsi in India, which included a new plastic recycling plant in India."[137] These efforts and communiqués demonstrate a high level of solidarity, networking, and coordinated action occurring between Indian and northern TSMOs around this campaign that exclusively targeted a transnational corporation (TNC).

Pepsi's marketing slogan used across India in the 1990s was, "Yehi Hai Right Choice, Baby!" which translates to, "This is the right choice, Baby!" Indian activists with PIRG declared that "Pepsi's slogan is quite appropriate in the present context. The way the Indian economy is opening up in the name of deregulation, liberalization and competitiveness under the structural adjustment program, there are enough indicators to demonstrate that India is becoming the 'Right Choice' for dumping wastes of the Northern countries."[138]

That same year, 1994, Pepsi had the unenviable honor of being placed on the *Multinational Monitor* magazine's infamous "Ten Worst Corporations" list. The main reason for this listing was Pepsi's investments in Burma, which activists critiqued and then called for a boycott against the company. Liz Claiborne, Levi Strauss & Co., Wal-Mart, Reebok, Eddie Bauer, and Amoco all decided not to do business in Burma, which has been run by a repressive junta for many years. When confronted with the threat of a boycott, Pepsi CEO Wayne Calloway accused human rights activists of "dealing in coercion and strong-arm tactics." He argued, "It's no different than what years ago was practiced by Joe McCarthy and the like."[139] This bad press gave the activists focused on the situation in India hope that perhaps the company would agree to make some changes.

Pepsi Responds

In a letter mailed from Pepsico Foods and Beverages public relations to people who participated in Global Response's letter writing campaign, manager Brad Shaw stated:

> Recently, several groups have engaged in a campaign that maligns and misrepresents Pepsi's business in India. The groups have claimed that Pepsi is dumping PET [plastic] waste in India and is subjecting its workers to unsafe conditions and unfair wages. None of this is true. We should note that Futura Industries . . . has no relationship with Pepsi, [and] *is* shipping used PET packaging to India, where we are told that it is fully recycled into fibers used in clothing, carpets, and other materials. Futura has assured us that each step in its recycling process—including the safety measures that it employs—meets or exceeds Indian government regulations. Additionally, we're informed that the wages earned by Futura employees meet all government standards and, in fact, far exceed India's per capita wage.[140]

Through this statement, Pepsi made it clear that it had no legal relationship with Futura. However, this is a common corporate tactic, whereby firms use middlemen businesses to create distance between themselves and firms in countries where unsavory practices may be occurring. This is similar to outsourcing and subcontracting, which allow for limited or no legal liability when public interest activists may seek restitution for alleged violations of laws or policies. Pepsi set up the PRCC, which served as a middleman firm that then did business with Futura. Ron Kemalyan, the broker for PRCC's Los Angeles office, admitted that despite his company's name (which includes the word *recycling*), it maintains no actual recycling facilities and only facilitates the export of the waste. He stated, "We are a brokerage."[141]

Brad Shaw offers a typical corporate perspective on recycling while taking credit for a milestone Pepsi reached: "In 1990, Pepsi introduced the first PET [polyethylene terephthalate] bottle using recycled plastic and, by next year, we're hoping to introduce a PET package that contains more than 50 percent recycled plastic. In short, we're working hard to better manage the waste stream and recycling is a big part of our effort. In our opinion, recycling is good for the environment and it makes good business sense."[142] In fact, recycling toxic products like plastic pollutes ecosystems, and it is certain that Pepsi pursued recycling less for its environmental value and more for the positive public relations and profit potential.

Years after this conflict broke out, the depths of corporate harm to India's environment and public health became even clearer. In 2003, one of India's leading NGOs, the Center for Science and Environment (CSE), released a report declaring that soft drinks manufactured in that nation, including Pepsi, contained high levels of pesticide and insecticide residues. The report was based on laboratory analyses of samples from twelve major soft drink manufacturers selling products in India and found that all of them contained residues of lindane, DDT, malathion, and chlorpyrifos—extremely toxic pesticides or insecticides. Officials at Pepsi and Coca-Cola issued emphatic denials, but, significantly, the CSE laboratories tested samples of soft drinks that are commonly sold in the United States as a control in their study and found that they did not contain any pesticide residue.[143] There appeared to be a double standard for quality that penalized Indian soft drink customers with toxic-laced products that U.S. consumers need not be concerned with.

As if toxic soft drinks and plastic bottle dumping were not enough, plans for a Pepsi plant in Kerala state, India, were temporarily put on hold in 2004 as officials canceled the license owing to concerns that the plant's use of groundwater to produce the soft drink would threaten the already overtaxed and scarce drinking water source in the area. Finally, in August 2006, the state of Kerala banned the production and sale of both Pepsi and Coca Cola because of continued worries over water scarcity and possible health effects of pesticides believed to be contained in the soft drinks.[144]

The Struggle Goes On

Ultimately this case also amounts to an indictment of much of what passes for recycling in the United States and elsewhere. Greenpeace found that more than half of the plastic waste exported by the United States to destinations in Asia comes from households, which is collected through various municipal and community recycling programs.[145] This practice renders the plastic waste out of sight, out of mind for U.S. consumers, who feel good about recycling because, according to common wisdom, recycling is "good for the environment." A Greenpeace study on plastics recycling in the early 1990s concluded: "It is increasingly likely that the plastic bags and bottles that you drop off at your local recycling center or are picked up on your curbside may end up in the countryside in China or an illegal waste importer's shop in Manila. The plastics industry is catching on to the tried-and-true practices of international waste traders worldwide. By exporting their wastes to less industrialized countries, they've learned they can avoid domestic regulations, avoid community opposition to waste handling facilities, pay their workers pennies a day, and maintain a 'green' image at home."[146]

That is, in the Pepsi case, ecological modernization in the United States is made possible through environmental injustice in Asia. Rather than reflective of ecologically sustainable development, this kind of recycling seems to be more representative of economically sustained development.[147] This case also demonstrated the stark geocultural hierarchies between the United States and India, where it was perfectly acceptable to the state and to Pepsi to send northern consumers' waste to a nation of people who are valued less because they are poor and non-European. As activists in the United States and India stated repeatedly, this was a form of transnational environmental injustice. Indian activist and researcher Nityanand Jayaraman explained, "There is an unwritten code that justifies development for some at the cost of others (usually, people of color or people from disadvantaged communities or classes). The sentiment that somebody has to pay for development is probably the most damaging and dehumanizing ethic that defines development today."[148] And while the U.S. and Indian governments facilitated this environmental injustice, it was a collection of corporations in the North and South that drove the process: Pepsi, the PRCC, and Futura. Private profit trumped the public good and environmental health in India, while attempting to preserve both in the United States. The power of transnational corporations remains unparalleled in world history. As Jayaraman put it, "Without exception, the impact of transnational corporations in India has been negative. In all the campaigns that I have been associated with, the industry has breathed down the government's neck, has lied, intimidated activists, and delayed action. Most of our work has been little but an ant's bite on the corporate backsides."[149]

The reality of the political economic opportunity structure is that the power of corporations combined with state institutions is often unbridled and quite brutal.

This case was never resolved, and today, India remains a favorite dumping ground for plastic wastes, mostly from northern nations such as the United States, Canada, Denmark, Germany, the United Kingdom, the Netherlands, Japan, and France. According to the Indian government, at least 59,000 tons and 61,000 tons of plastic wastes were imported in 1999 and 2000, respectively.[150]

Many activists in India and other parts of South Asia are hopeful and see the possibility for reversing the tide of environmental injustice. For them, resisting the trade in toxic wastes is much broader than an environmental concern. "South Asia is a region that has suffered from great disunity in the past," explains Farhad Mazhar, managing director of UBINIG, a progressive policy research and advocacy organization in Bangladesh. He continued, "A campaign to end the import of toxic wastes provides the opportunity for all countries, all political parties, all communities, in fact all South Asians, to come together for our mutual benefit. By declaring South Asia a toxics free zone, we will also be laying the groundwork for increased regional unity on many other issues. At the same time, we realize that shifting the toxics to another region is not the solution, so we will work with other regional campaigns to free not only South Asia, but the whole world, from toxic wastes."[151]

The Pepsico case reveals some fascinating complexities in transnational hazardous waste dumping that speak to the concept of "delayed dumping" because it shows that these nations are not just being dumped on, rather they are also using these wastes as a path toward modernization.[152] That is, the majority of the plastic that Pepsico dumps in India is recycled and turned into bottles for sale to Pepsi, generating some local jobs and income. Other forms of transnational environmental inequality reveal this pattern as well. Electronic wastes in China, for example, are recycled into computers and other electronics goods for sale in global consumer markets; the technology transfer of incinerators from North to South is used to incinerate waste exported to the South and waste that consumers and industry in the South produce themselves (thus, incineration is viewed as a way of controlling environmental problems but also as a method of modernizing a nation's urban waste management systems); and the transfer of toxic pesticides and fertilizers from North to South was hailed as a "Green Revolution," a massive effort to modernize the Third World's agricultural systems. If this is the good news about technology transfer, then the global South really has a dilemma, and institutions that embrace and push this model are blind or unfeeling to the myriad negative social and environmental consequences that follow.

Conclusion: Back to the Global Scale

Although each of the case studies is instructive for understanding the scourge of global environmental inequality, only recently has there developed a set of more formal international networks that link these struggles across national borders (see chapter 3). Greenpeace, the first global SMO to tackle these problems, was also often criticized for ignoring the need to engage in movement building, for being "an organization in a hurry."[153] Greenpeace was known for putting out fires and "parachuting" activists into hot spots around the world where urgent action was needed to stop destructive waste management projects, but the long-term commitment to local organizing was rarely there. GAIA stepped in to fill that void when it formed in 2000 at a meeting of activists in Johannesburg. GAIA now works to facilitate local, community-based grassroots organizing in dozens in countries around the globe. It opposes incineration projects and promotes zero waste, recycling, reuse, and composting. Each year GAIA coordinates a Global Day of Action Against Waste, when activists in scores of countries take on an issue important to their local and domestic campaigns as it relates to waste and publicize their work or the threats they are confronting. The first Global Day of Action Against Waste was in 2002, during which 126 groups in 54 nations took part, and each year it grows. The GAIA secretariat is based in the Philippines and issues simultaneous press releases from Manila, the United States, Buenos Aires, and many other cities, presenting information on the actions to media across the globe. In the run-up to the Global Day of Action in 2004, the GAIA secretariat sent a message to the International Campaign for Responsible Technology (ICRT), itself a global environmental justice network focused on the electronics industry (see chapters 3 and 6), that reads as follows:

> Dear Friends at ICRT: Warm greetings from the GAIA Secretariat. We write once again to urge you to get involved in the GAIA-organized Global Day of Action on Waste on September 1, 2004. As of today, 80 groups and individuals from 33 countries from Azerbaijan to Zimbabwe have confirmed their intent to take part. Why is your participation very important? The following letter from Peter Lach-Newinsky of the Residents Against Dioxins (RAD) in the Southern Highlands, Australia gives good reason for you to participate: "We think both the Global Day of Action on Waste and GAIA itself a very good idea for global networking and awareness-raising. It is good to see citizens all over the world communicating and learning from each other as they fight similar struggles. We need to be as global in our resistance as our adversaries are in their capital-driven attempts to trash the planet and our health for short-term shareholder gain."[154]

Across the globe, through local and globally coordinated efforts by GAIA and its member organizations, popular resistance efforts have stopped incinerator proposals, shut down existing incinerators, and worked to pass legislation to ban or restrict waste incineration altogether. This kind of action is ongoing around the world. In 2001, for example, community opposition forces defeated major waste incineration proposals in France, Haiti, Ireland, Poland, South Africa, Thailand, the United Kingdom, and Venezuela. Global environmental justice movement networks rooted in both local grassroots power and transnational coordination represent one of the best hopes for pushing states and industry toward more socially and ecologically sensible policies. This kind of action will be needed more than ever before as the incinerator industry has seen fit to push its products on the global South (including Central and Eastern Europe), including not only municipal waste but also medical waste incineration, which has received support from institutions like the World Bank.[155] GAIA teamed up with TSMOs like Health Care Without Harm (HCWH), and hundreds of grassroots SMOs around the globe to meet this challenge. HCWH, serious about zero waste and environmental justice, published a tool kit for activists titled *How to Shut Down an Incinerator.* In that document, under the section titled "Building an Anti-Incinerator Campaign," there is a discussion about direct action:

> An action is any step you take to advance your group's goals. Petitions, letter-writing campaigns, and educational meetings are all actions that advance your

group's goals. A direct action is the more dramatic type of action, involving confrontation and demands. Direct action begins after your efforts at education, information-sharing and persuasion are ignored. Use direct action when your group is ready to confront a decision-maker with its frustrations and to make specific demands. Direct actions move your organization *outside the established rules* for meetings and discussion. It takes your group into a forum in which you make the rules and where *elected representatives and corporate executives* are less sure of themselves and how to handle the situation. A direct action often provides the necessary pressure to force your target to act on your group's issue.[156]

This document reveals how TSMOs are truly focused on movement building and empowering communities to secure environmental justice in the face of the powerful state-corporate nexus I call the political economic opportunity structure.

The way that TSMOs like GAIA and HCWH function is quite remarkable. They insist on not dominating local struggles, only supporting them. They also have moved beyond the typical problem focus of many social movements toward implementing and embracing solutions. For example, GAIA works closely with waste pickers (or scavengers) in the informal sector in many nations, who seek recognition and protection of their livelihood, since they contribute to waste recovery and reuse of valuable materials for communities. GAIA notes that waste pickers are an essential component to most nations' waste management systems, and their generally degraded employment conditions must therefore be improved. One example is in Cairo, Egypt, where the scavengers, or the *zabbaleen,* have organized themselves into an advocacy organization that oversees a waste collection and recycling system that diverts 85 percent of collected waste and employs 40,000 people. The zabbaleen have consciously rejected the label of "scavenger" and have insisted they are "resource recovery workers."[157] GAIA has worked to support the zabbaleen efforts and those of similar operations in many countries in the South. They connect the local struggles of communities against harmful waste management practices by corporations and states and in favor of socially, economically, and environmentally progressive projects across the globe. This is the global environmental justice movement at its best.

Transnational, regional, national, and local environmental justice activists and networks opposing garbage dumping, sham recycling, and incineration are confronting the seminal environmental justice issue: garbage. Societies have always produced waste and have always required places to site that waste. The detritus will always flow along the path of least resistance, but when vulnerable and targeted communities challenge that social order, they can rewrite those rules. Activists involved in these struggles in Haiti, Philadelphia, the Philippines, and India were aware that toxics were being transferred from one part of the world to another, through what I term the hyperspatiality of risk (the problem of shifting waste flows from one space to another, as a result of political opposition or economic opportunity). They were aware that economic, political, and cultural hierarchies provided environmental privileges for some and pain and disadvantage for others; aware that this process is rooted in both racial and class inequalities and in differences in how nature and the environment are imagined in the minds of those controlling powerful institutions producing these inequalities; and aware that those institutions that must be confronted include and extend beyond the state, to corporations and even environmental organizations. What they may not have been aware of was how powerful the global movement for environmental justice had become.

5

Ghosts of the Green Revolution: Pesticides Poison the Global South

Pesticides are an ideal product: like heroin, they promise paradise and deliver addiction.

—Paul Erhlich[1]

Paraquat is very dangerous and today I know it is a highly toxic pesticide. A poison is a poison. It is made to kill.

—Arjunan Ramasamy, fifty-six-year-old plantation worker in Malaysia[2]

Food, one of the very foundations of culture—to say nothing of sustenance—is imbued with tradition and symbolic, linguistic, and historical significance. Today food—both agricultural produce and livestock—is heavily laden with pesticides and a range of persistent organic pollutants (POPs)—toxic chemicals that remain in ecosystems for lengthy periods of time, travel long distances, and accumulate in the food chain (see table 5.1). Like global environmental inequality, POPs are a critical hallmark of the risk society in a late modern nation-state and global economic system.

The systematic production, utilization, and export of POPs and pesticides began in the post–World War II era but grew exponentially during and after the Green Revolution of the 1970s.[3] This was a project of global North nations and agrochemical firms that engaged in the technology transfer of modern, or industrial, agricultural production methods to southern nations. One of the axes of this project was the belief that heavy pesticide and fertilizer use would lead to greater efficiencies by producing larger crop yields, particularly in nations and regions with widespread poverty and hunger. This assumption was problematic because this model of modernization ignored the devastating public health and ecological

Table 5.1
Pesticide Action Network's "dirty dozen" list

DDT
2,4,5-T
Aldrin (including dieldrin and endrin)
Camphechlor
Chlordimeform
Chlordane (heptachlor)
DBCP
Ethylene dibromide
Lindane
Parathion
Paraquat
Pentachlorophenol

Source: www.panna.org

harm associated with such toxic inputs.[4] The export and utilization of countless tons of pesticides in Africa, Asia, the Caribbean, and Latin America has contaminated watersheds, communities, and human bodies with unprecedented reach and destructive effect. Perhaps most ironic, pesticides have rendered highly toxic and hazardous the very land where many crops are grown. This is how the notion of modernization has become inextricably linked to the theory of the risk society because these development schemes have exported chemical and environmental hazards to the nations they were intended to support. And although everyone on the planet ultimately is affected by this chemical scourge, this risk is racialized, classed, gendered, and imposed differentially across geographies and communities for political, social and cultural, and economic reasons.

Through the presentation and analysis of trends and case studies in this chapter, I consider the origins and impacts of the Green Revolution on global South communities. I also focus on the ways in which SMOs and activists in the global South and their partners in North America and Europe are developing new knowledge and resistance practices to combat the transfer of pesticide hazards to the South while developing sustainable and socially equitable solutions for their communities. These activists are carrying on the next generation of the struggle against pesticides and the human exploitation that accompanies their use that the great farm labor organizer Cesar Chavez and the great activist-scientist Rachel Carson began in the 1960s.[5]

The Circle of Poison

What is distinct about pesticides, when compared to other kinds of pollution and waste, is that they are produced for the expressed purpose of killing something, of ending the lives of plants or insects deemed to be invasive or alien pests or weeds in an agricultural context.[6] This is all intended to allow the unfettered production of food—a life-giving substance—but this goal is significantly undermined by the contradiction embodied in killing or poisoning elements of the environment (the soil, water sources, and flora and fauna) that make food production possible. Chemical wastes, garbage, electronic wastes, and other forms of pollution discussed in this book are distinct from pesticides because they are the externalities, the unwanted by-products of production. Pesticides, on the other hand, *are* the product being produced and sold. They are created to enable production through the death of significant components of nature.[7] Thus, through pesticide application, we are at once engaged in imagining the environment as a life-giving resource *and* as a force with which we are at war. This logic dictates that we must control and do violence to the ecosystem in order to make it work for us. This ideological framework is similar to those that support systems of social domination (slavery, militarism, patriarchy, classism, racism, and others) as well. The practice of harming the ecosystem around us in order to produce food for life-sustaining purposes should be cause for concern. The life-harming purpose of pesticides is perhaps most disturbingly obvious in the thousands of suicides by pesticide ingestion over the past several years by farmers in numerous countries in the South who find themselves locked in hopeless spirals of debt to international banks.[8]

Although pesticides use underwent a transformation in quality and quantity after World War II, they are not a new phenomenon. Ancient Romans burned sulfur to kill insect pests and used salt to control unwanted weeds. In other places, ant controls were introduced through a mixture of arsenic and honey during the seventeenth century, and by the late 1800s, farmers in the United States sought to control unwanted insect species through the use of sulfur, nicotine sulfate, calcium arsentate, and copper acetoarsenite (also called Paris green).[9] U.S farmers in the nineteenth century used lead arsenate on apples and grapes to ward off pests. In the 1940s, DDT and related chemical substances came into use in agriculture and as a public health measure for malaria control. These chemicals were far more toxic than those used in previous generations, and the volume used increased dramatically.[10]

The U.S. Environmental Protection Agency (EPA) banned DDT in 1972 after an abundance of evidence came to light concerning its harmful effects on ecosystems. Today most pesticides banned for use in the United States are exported, dumped, or used throughout the global South—a clear case of global environmental inequality and racism.[11] While one might like to think this is a case of out-of-sight, out-of-mind hazard transfer, even the most callous of U.S. consumers should be concerned, because when they eat fruits and vegetables imported from southern nations, these products are generally grown and sprayed with pesticides, thus exposing U.S. consumers on a routine basis.[12] This is what activists have called the circle of poison and reflects the essence of Beck's risk society approach, Guinier and Torres's miner's canary, and my racism and classism as toxicity frameworks, because everyone is affected. Even so, from production to consumption and disposal, farmworkers, the poor, women, children, indigenous peoples, immigrants, and people of color in the United States and globally tend to be affected more heavily by pesticides, so the risks are borne unevenly. As Rafael Mariano, the national chairperson of the Peasant Movement of the Philippines, stated at a conference on pesticides and justice, "The Third World accounts for 99 percent of deaths from pesticides."[13] Annually, there are more than 25 million reported cases of pesticide poisoning in the global South alone.[14] Within the global North, environmental inequalities from pesticides prevail as well. In the United States, the rise in consumption of pesticide-laden foods over the past several decades, combined with weak labor laws and racial discrimination, produces a situation where farmworkers and their families confront abnormally high rates of pesticide-related illness, poisoning, cancer, and birth defects,[15] and consumers, particularly children, may be placed at risk from eating pesticide-laden foods.[16] Environmental justice efforts to protect workers, communities, consumers, and ecosystems from pesticide poisoning require a global approach, because these toxics pervade every living being on earth and because the global companies controlling the research, production, and distribution of these poisons are formidable political-economic forces, with easy access to congress, parliaments, legislatures, and heads of state, and with annual sales of more than US$30 billion.[17]

Social Inequality, Labor, and the Ecology of Pesticides

Indigenous Peoples and Women From production, to use, and finally to disposal, pesticides destroy ecosystem life and harm the most vulnerable human communities. Women, children, migrant farmworkers, and poor populations suffer especially great burdens from pesticides. Not surprisingly, indigenous peoples do as well. For example, Plan Colombia's aerial spraying of tons of pesticides on coca and poppy plants in Colombia disproportionately harms indigenous peoples' land and health in that nation, violating their right to a clean environment, as is customary under international law.[18] Indigenous communities around the globe are also disproportionately harmed by POPs, which include a number of pesticides that travel great distances through air and water systems. Traditional food sources used by Arctic indigenous peoples, for example, are exposed to POPs and heavy metals produced by northern nations.

While farm labor is not easy for men, it has placed a major burden on many women throughout the world. Although women constitute a majority of the world agricultural workforce, they are frequently excluded from decision making around crop and pest management, they tend to be concentrated in gender-specific jobs, they earn lower wages than men working in agriculture, and they are exposed to pesticide poisoning with greater frequency. These forces converge to create greater risks to women's physical and reproductive health and the health of their children.[19]

When workers raise concerns about agrochemicals, women are frequently unable to attend meetings of farmers when experts educate them on the risks associated with pesticide use. Instead, they are often working in the fields, providing child care, or preparing food in the home. Women also have little control over which pesticides are used, because landowners and pesticide dealers typically make those decisions. Female agricultural workers in the South are often illiterate and therefore unable to read and understand the labels on pesticide containers, which might otherwise reveal safety protocols as well as risk factors to the user.[20] In many nations, rural women often reuse pesticide containers for storing or transporting crops.

Women face multiple barriers to their safety when working in areas where pesticides are used. Not only do they generally face greater exposure to pesticides than do men, they are frequently provided with little or no information, training, or safety equipment on the job, and they face great difficulties accessing medical care for the effects of the pesticides. A study of female agricultural workers in Southeast Asia reported: "When these women from the plantations, with their particular cultural sensitivities, are faced with often hostile and indifferent male medical staff who do not speak their language, they are intimidated into silence or frustration at not being able to make their illnesses understood. This lack of cultural sensitivity does not ensure women's health is being protected."[21]

Medical studies have revealed associations between pesticides exposure and a range of reproductive health ailments, including stillbirths, miscarriages, birth defects, infertility, spontaneous abortions, and delayed pregnancies.[22] Recent medical research also finds associations between pesticide exposure and breast cancer.[23] These problems come in addition to the routine health effects of pesticide exposure that can include itching, nausea, dizziness, muscular pain, sore eyes, nervous system damage, skin diseases, nosebleeds, fingernail removal, and breathing difficulties.[24]

The power of privatization and transnational corporations (TNCs) plays a significant role in this story. The introduction of global corporate agricultural firms tends to increase women's dependency on external resources controlled by a small group of companies. TNC-controlled agriculture also increases women's risk of exposure to pesticides and genetic engineering, all of which can contribute to women's health risks and impoverishment through low wages and displacement from farm ownership.[25] Despite these pressures, women are resisting such exploitation, fighting against the construction of dams, occupying lands threatened with external corporate seizure, and working to repel TNCs that farmers' associations have labeled dangerous and harmful to local food production and local cultures. Many women are also practicing sustainable agriculture at the community level.[26] The topic of gendered impacts of pesticides is closely related to the issue of the effects these chemicals have on the general health of the global agricultural labor force.

Occupational Safety and Health Issues and Workers' Rights The International Labor Organization estimates that in 1994, there were 2 to 5 million occupational cases of pesticide poisoning, with 40,000 resulting in fatalities.[27] In some nations, pesticide exposure is believed to be the cause of 14 percent of all occupational injuries in the agricultural sector and 10 percent of overall fatalities.[28] Many studies conclude that farmers have a greater likelihood of developing brain, skin, and prostate cancers, leukemia, non-Hodgkin's lymphoma, and other ailments than the general population. Their employees—farmworkers—are at even greater risk because of the more direct exposure to pesticides that comes from application.[29] One study of Latino farmworkers in California revealed that when compared with the larger Latino population in the United States, they had a greater probability of developing leukemia (59 percent), cervical cancer (63 percent), uterine cancer (68 percent), and stomach cancer (70 percent).[30] Finally, the health effects of pesticide exposure are often long lasting. Research demonstrates that farmers who have used agrochemicals on their crops in the past experience continued and intensified neurological disorders long after they ceased using such products.[31]

The health risks associated with pesticide exposure "tend to fall on those benefiting the least from pesticide use. . . . This is a pattern reported for various LDCs [lesser developed countries] in Africa, Asia, and Latin America."[32] These trends suggest that environmental inequalities associated with pesticide use are widespread and observable across race, class, gender, age, migration status, and nation.

Environmental Impacts of Pesticides The effects of pesticides on human health are, in many cases, devastating, but the damage done to ecosystems is almost immeasurable in its immensity and reach. As the Pesticide Action Network Asia Pacific argues, pesticides poison "our bodies, our wombs, our children . . . our water, air, soil, and our food."[33] Widespread contamination of ecosystems, water sources, and agricultural land is associated with the application of these chemicals in many countries.[34] Toxics Link India, an activist nongovernmental organization (NGO) working on environmental justice issues in New Delhi, declared: "The benefits of the Green Revolution in India have long worn off and have left in its wake contaminated soil, water, air, and food. There is no 'perfect pesticide' which kills only pests and is non-toxic to everything else. The use of pesticides affects all kinds of life resulting in environmental imbalance."[35]

Perhaps more than any other kind of pollutant, activists link pesticide production and its negative impacts directly to the efforts of TNCs to prioritize profit and shareholder value over public health and the environment. TNCs bear a great deal of responsibility in the political economic opportunity structure that environmental justice activists engage because they are the institutional embodiment of the idea of racism and class domination as toxicity.

Thus, while pesticide applications may indeed help protect some agricultural settings from certain invasive plant species, there are great human health and environmental costs associated with their use, and global environmental justice activists blame a relatively small group of transnational firms for some of the worst abuses associated with manufacturing and marketing these products.

Economic Globalization, Environmental Injustice, and the Violence of Markets

Economic globalization entails erasing barriers to trade across national borders, so that investors and industries can more easily access markets and production sites where costs, regulations, and other obstacles may be much lower and lax than in other nations. This is why some northern businesses seek migrant workers who move North and why other northern corporations move production facilities to the South. This dynamic offers southern nations on a path toward modernization little incentive to improve their own regulatory climate. Thus, workers, activists, and communities in the South have seen an enormous increase in corporate-led environmental threats and wage instability since the 1970s, when globalization really took hold by way of TNC expansion southward through trade liberalization and the Green Revolution.[36]

Globalization often means that cultures and nations are more tightly integrated than ever before. Unfortunately for farmworkers, who are paid very little and generally not allowed to organize into labor unions, this integration frequently leads to further disempowerment, as foreign and domestic TNCs make decisions that negatively affect their everyday lives.

One result of this process is the crossing of borders by workers and pesticides in ways that leave the former at a linguistic disadvantage. For example, on many farms in Cambodia, storage containers with the highly toxic methyl parathion pesticide carry safety and warning labels in Thai rather than the Khmer language, so Cambodian farmers are unable to read the instructions for use and any warnings they may carry.[37] The same problem is documented in Malaysia, where migrant women workers from India are unable to read safety warnings and labels on pesticides on the plantations where they work because they are printed in another language.[38] So although globalization might mean greater integration of markets and more frequent cultural crossings, such interactions often result in heightened risks being borne by populations with less political, cultural, and economic leverage.[39] Migrant and agricultural workers are generally squeezed by the deep inequalities that globalization creates by design.

The heightened toxic exposure that farmers, farmworkers, and surrounding communities face from pesticides is linked to the simple fact of global environmental inequality. Although most pesticides are manufactured and used in the global North, southern nations are substantially increasing their use of these substances as they shift from North to South: "Between 1972 and 1985, pesticide imports grew by 261% in Asia, 95% in Africa, and 48% in Latin America. In 1991 LDCs accounted for over 30% of total world pesticide consumption, whereas the comparable figure for 1985 was near 22%."[40]

Pesticide exports to the global South have grown in volume in recent years, according to a study by the Foundation for Advancements in Science and Education (FASE). Between 1997 and 2000, nearly 3.2 billion pounds of pesticide products were exported from the United States, which represents a 15 percent increase over the 1992–1996 period. Of that 3.2 billion pounds, nearly 65 million pounds of those pesticides were already banned or severely restricted in the United States. The vast majority of these materials were sent to southern nations.[41] FASE also found that 9.4 million pounds of never-registered pesticides were exported in 1995 and 1996, a 40 percent increase since the period 1992–1994. During the late 1990s, the United States exported in excess of 28 million pounds of pesticides officially deemed extremely hazardous by the World Health Organization, which amounts to a 500 percent increase since 1992.[42] When substances deemed to be hazardous are no longer used in the United States but are exported by U.S. companies to poor nations, this raises troubling questions.

In 2001, environmental justice advocates received support for their claims of global environmental inequality in the case of pesticide exports when UN special rapporteur Fatma Zora Ouhachi-Vesely made public statements critical of the U.S. practice of exporting to southern nations chemicals, pesticides, and waste that were banned domestically: "Just because something is not illegal, it may still be immoral. Allowing the export of products recognized to be harmful is immoral. . . . Developing countries do not have the medical or regulatory capacity to address the negative effects of these chemicals on their population. That is what makes this an immoral practice."[43]

The race and class exploitation of the global pesticide trade become clear when a double standard is applied to those toxics exported to other nations. For example, around 30 percent of pesticides marketed in the global South fail to meet internationally accepted standards of quality set by the United Nations Food and Agriculture Organization (FAO), yet they remain on the market. The problem is most acute in Africa, where quality control efforts are severely lacking.[44] Furthermore, according to the FAO, pesticide makers in the United States often fail to warn purchasers and users overseas about the hazardous nature of these substances. As Carl Smith, director of the FASE pesticide project, noted:

> There is a real problem in Latin America with restricted-use products. . . . Nearly two-thirds of these countries reported that the labels on pesticide products do not contain the same recommendations for safe use that would be expected in the U.S. . . . This whole subject hits a nerve because it raises not only trade issues, but also moral issues. . . . We're talking about chemicals that are so dangerous that the U.S. has either banned them outright, or made it illegal to use them without a full-body protective suit and a respirator. How do you justify sending them to places where they'll end up being sprayed by shirtless farmworkers, often children, who have no idea of their hazards?[45]

The causes of global environmental inequalities in pesticide use are varied and complex. External economic forces play a heavy hand in pushing pesticides on the South. Since the early 1980s, the International Monetary Fund (IMF) and World Bank have required global South nations to implement structural adjustment programs if they desired economic aid, investment, or debt relief. Such arrangements often bring a large volume of pesticide use to these nations because the export-oriented agricultural framework of most international loan programs is built around the production of crops like bananas and cotton, which require more pesticides than do food crops for domestic consumption.[46] These agreements provide new and growing markets for agrochemical corporations, particularly since the human health and environmental impacts of pesticide pollution are becoming known in the North, reducing their popularity and introducing new restrictions and bans on certain chemicals in the United States and Europe. Between 1988 and 1995, the World Bank financed the purchase of $250.8 million worth of pesticides, mostly from agrochemical TNCs in northern nations, which were then sent to southern nations as part of international development aid packages.[47]

Some scholars describe the flow of pesticides through the world economy as a "system of dependent relations between countries holding dominant and subordinate positions."[48] Several forces converge to produce this relationship, including widespread poverty among southern nations; loans from international financial institutions (IFIs) such as the World Bank, the IMF, the Export/Import Bank, and the Overseas Private Investment Corporation; domestic government subsidies and agricultural policies that support the purchase of, for example, machinery, hybrid seeds, irrigation, fertilizers, and pesticides; and aid packages from private foundations (such as the Rockefeller and Ford foundations) and international aid and development associations (USAID and the FAO) that encourage pesticide use in the South.

Another critical process in play is the growing regulation, including bans or heavy restrictions, of pesticides in northern nations as a result of the efforts of environmentalists and farmworker advocates who have brought attention to the widespread social and ecological harm these substances cause. As with other environmental and labor dynamics in a global economy, such regulation often does not solve the problem and instead provides TNCs with incentives to seek friendlier business climates. For example, after the 1973 ban on the chemical DDT in the United States, TNCs responded by looking to foreign markets to sell pesticides, and they achieved unprecedented success. This ban on DDT and increased regulation of other pesticides and chemicals (as a result of the emergence of the U.S.

Environmental Protection Agency and the Occupational Safety and Health Administration) happened to coincide with the Green Revolution of the 1960s and 1970s, which allowed TNCs making pesticides to essentially shift markets southward. The late 1960s and early 1970s was a time when the environmental movement in the North succeeded in creating an increased environmental awareness among the citizenry, which also led to a rise in the growth of organic farming, both of which facilitated a shift of pesticide use southward.

The race and class inequalities embedded in these geographic and spatial hierarchies are revealing. Pesticide marketing is controlled by a small group of TNCs based in northern nations. In fact, in 2002, just six companies—Syngenta, Monsanto, Dow, Bayer, BASF, and DuPont—controlled 80 percent of the pesticide market.[49] This process is fueled by the desire for economic growth (on the part of TNCs and southern nations seeking development and loan forgiveness) and race and class inequalities between and within northern and southern nations. As a result, ethnic minorities, migrant workers, women, working-class persons, farmers, and indigenous peoples are being hurt the most by the proliferation of pesticides in the South, while corporations enjoy healthy profits. Hence, it is not surprising that the bulk of activism around pesticide concerns is generated by groups supporting these vulnerable populations, who then focus their grievances on TNCs and IFIs.

The Green Revolution was the North's answer to the problems of poverty and hunger in the South. The adoption of chemical-intensive, large-scale, export-oriented monoculture was intended to increase local production to feed the hungry, allow debt repayment, and increase economic competitiveness. Unfortunately, poverty, hunger, and indebtedness remain widespread throughout the South, while pesticide firms enjoy large profit margins.

The pesticide trade is often caught up in the broader scope of the illegal—or barely legal—toxic waste trade, wherein firms from northern nations dump outdated or banned pesticides and chemicals in southern nations.[50] This matrix of activities amounts to structural and cultural violence committed against people and nature. However, what many people may be unaware of are the links between pesticides and more formal practices of violence.

War and Pesticides

Anyone who seeks evidence that pesticides and insecticides are agents of death need look no further than the battlefields of the twentieth and twenty-first centuries, where these chemicals have been used to inflict great suffering on combatants, civilian populations, and entire nations. Indeed, the history of synthetic pesticides begins not with agriculture and efforts to protect and produce more food for the good of humanity but with warfare and the mass slaughter of human beings. "Synthetic pesticides were first developed as weapons and tools of war in World War II. . . . Pesticides and war are indiscriminately destructive and needlessly wasteful of human life and nature."[51] The following examples should sufficiently illustrate this point:

• During World War I, the Bayer Corporation developed poisons, including mustard and chlorine gases, that were used in trench warfare. As part of the larger IG Farben Company, Bayer developed the insecticide Tabun in 1936 under the direction of Gerhard Schrader. Schrader also later discovered the toxic nerve agents Sarin and Soman, as well as the chemical compound E 605, the principal ingredient in the pesticide parathion. IG Farben was dissolved after the war, but Schrader stayed on to work for Bayer. Soon after World War II, Bayer and other firms released a range of organophosphorus compounds, including parathion, into the marketplace as insecticides for agricultural use.

• Dow Chemical Corporation manufactured the infamous chemical herbicide Agent Orange, which is widely viewed as having played a central role in one of the worst cases of toxic warfare and environmental racism in history: the U.S.-Vietnam War. During this war, the U.S. military employed Agent Orange to destroy foliage that offered Vietcong soldiers camouflage. Agent Orange has extraordinarily high concentrations of dioxin, the most toxic substance known to science, and has caused irreparable harm to the health of U.S and Vietnamese soldiers and civilians, as well as Vietnam's ecosystem. Operation Ranch Hand was the name given to the military campaign involving the dumping of an estimated 12 million gallons of Agent Orange and other chemicals on South Vietnam, causing physical deformities in tens of thousands of children and destroying 14 percent of that nation's forests. The Vietnamese government recently reported that more than 70,000 of its citizens suffer from medical diseases related to Agent Orange exposure. Other estimates are closer to 1 million because many people who suffer were not born at the time.[52] As one researcher writes, "Today the Vietnam 'war' is still being waged. The military hostilities are long gone, of course, but some 'battles' still rage daily and hourly. These battles are fought quietly, individually—mostly in the blood and body tissue of too many living things—in Vietnam, America, and other far-flung places seemingly unconnected to the war."[53]

• The U.S.-funded Plan Colombia and Colombia's "War on Drugs" involve not just supporting the Colombian government's military assault on a guerrilla movement, but the aerial spraying of tons of toxic herbicides to kill coca, cannabis, and opium poppy crops. This plan amounts to a massive transfer—or dumping—of hazardous chemicals from North to South, which threatens human health, the ecosystem, and the economy of that nation.[54]

Since World War I, the growth and prestige of the chemical industry has been tightly linked to warfare, as industry leaders and public officials in northern nations have frequently and successfully framed that sector as critical to national defense.[55] Pesticides and militarism are intimately linked.

International Agreements on Pesticide Production and Export

In an effort to ban the export of certain pesticides (including chlordane and heptachlor) that are illegal in the United States, the U.S. Congress tried to pass a "circle of poison" law outlawing this practice. The bills were nearly successful but were killed at the last moment when leaders from both chambers yielded to the first Bush administration's lobbying on behalf of agrochemical corporations. Even a much weaker version of this bill failed to pass in 1994 when the Clinton administration introduced it. Thus, there is no law in the United States against exporting and dumping banned pesticides, making global environmental injustice a de facto policy.[56]

Internationally, environmental regulations that do exist are minimal and largely weak, particularly in the global South. However, there are international conventions that can serve as guides for more reasonable pesticide regulation. Some examples follow:

- The International Code of Conduct on the Distribution and Use of Pesticides (1985) was adopted by the UN FAO in order to create rules and guidelines for the regulation, trade, and use of pesticides. This code serves as the globally accepted standard for pesticide management. It was revised in 2002 for the purpose of placing more weight on the reduction of risks and hazards associated with pesticides, particularly in the global South. The code encourages integrated pest management (IPM) methods for agricultural management and supports efforts to increase the participation of farmers (including women's groups) in this process (as opposed to the usual stakeholders controlling the crop management agenda: agrochemical firms and IFIs).
- The Stockholm Convention on Persistent Organic Pollutants (2001) bans the use of twelve POPs—toxic chemicals that persist in the environment for extended periods of time and migrate great distances and accumulate in the food chain. These POPs include nine pesticides (such as DDT and aldrin/dieldrin) and three toxic chemicals produced in incineration processes (dioxins, furans, and polychlorinated biphenyls). The convention obligates all participating nations to eliminate or restrict the manufacture, use, and trade of POPs. The United States still has not ratified the agreement, and the George W. Bush administration is vehemently opposed to it.[57]
- The Rotterdam Convention (2003) is an international treaty requiring nations that export hazardous materials to obtain the prior informed consent of importing countries before such transactions can take place. Importing countries are provided the tools needed to identify hazardous substances and reject them if need be, or manage them safely if they are imported. The convention covers at least twenty-two pesticides and five industrial chemicals, and more are to be added in the future. The major push for the signing of the convention came from environmental NGOs and consumer groups.[58]

Despite the efforts of transnational agrochemical firms and some U.S. officials, these are international conventions governing the use of pesticides that have the potential to offer greater protection to communities and ecosystems. The struggle to develop and implement these conventions and codes involves the efforts of grassroots activists and several transnational environmental justice networks.[59] In the rest of this chapter, I consider cases of local, regional, national, and transnational environmental justice activism designed to clean up abandoned pesticides and prevent further dumping by corporations and governments on communities in Africa, Asia, and the Caribbean.

Poison, Power, and Politics: Resistance against Pesticides

Although the actions of TNCs and governments are poisoning workers laboring in the fields of scores of nations, there is an active resistance movement demanding accountability, compensation, and environmental justice as people push back against agrochemical corporate environmental injustice. Two case studies address these struggles.

Obsolete Pesticides in the Bahamas

In 1978, U.S.-based Owens-Illinois Corporation sold all of its property and assets in one Bahamian island, Abaco, to the Commonwealth of the Bahamas. Until that time, the company had been the largest Abaco employer. Owens-Illinois owned and operated a sugar mill and plantation that, upon its exit, left a dearth of jobs in its wake. Abaco's economy and employment rate were harmed for many years afterward. Compounding these economic concerns were environmental threats no one anticipated. Left behind on the property were "more than 40 fifty-gallon drums of pesticides laced with highly toxic dioxin" as well as "the herbicide mixture of 2,4–D and 2,4,5–T, also known as Agent Orange."[60] Although the sugar mill was not located directly near a large population, it was above the water table that served as the main drinking source for most of the island's residents. A number of the drums were corroded and leaking chemicals into the ground, causing great concern among the island's citizenry when they discovered these materials years after Owens-Illinois departed.

A document from the Bahamian Ministry of Health listed the items contained at the sugar mill site, which included a "50/50 mixture of 2,4D and 2,4,5–T N-Butyl ester used as defoliant and pesticide (agent orange or dioxin) . . . isopropyl ester of 2,4D, weed killer . . . atrizine herbicide, ortho X."[61] The document continues, "The chemicals have been stored at the site since 1970. Some of the chemical barrels have labels. The site was initially used as a sugar mill when the chemicals were imported. . . . Tests done on October 16, 1989 by Flowers Chemical Laboratories Inc., confirmed the presence of 2,4D and 2,4,5T in several of the herbicides."[62]

The report confirmed the presence of dioxin at the site. Not only is dioxin chemically related to Agent Orange, it is also the most toxic substance known to science. In addition to these pesticides and herbicides, the site contained other hazardous materials, including asbestos and waste oils, which added to the general anxiety of local residents.[63] This case quickly became an ecological and public health crisis.

In 1991, a group of concerned citizens approached the Bahamian government to request that the site be cleaned up. reEarth, a Bahamian environmental watchdog organization, led the effort. Glen Archer, director of the government's Department of Environmental Services, declared that when the Owens-Illinois company sold the farm, the legal framework that would impose responsibility for any toxic waste left behind was not in place, particularly as it concerned a U.S.-based company operating abroad. In 1993, the Bahamas signed the Basel Convention, which made possible the assignment and enforcement of such responsibility.[64] Aside from the earlier legal conundrum this case posed, there were technical limitations that had to be addressed. Activist Sam Duncombe, of the environmental group reEarth, told her fellow activists, "The Bahamas does not have the technology or the expertise to deal with this kind of poison."[65]

This case also raised critical questions among local residents about U.S. politics and the view of the Bahamas from its northern neighbor. For example, a Bahamian journalist hinted at the double standard of global environmental inequality when he pointed out that dioxin "has been banned in the United States where areas in which it was used had to be evacuated."[66] That is, if the United States banned the substance and evacuated persons who may have been exposed to it, why should the Bahamas have to play host to the same pollutant?

At a public meeting held shortly after the issue became public, Department of Environmental Services director Glen Archer announced his office's plans to incinerate and entomb the pesticides on the island. The Abaconians at the meeting "flatly refused" to allow the plan.[67] As Sam Duncombe stated, "The threat of contamination by this dioxin could be the most serious environmental threat in The Bahamas."[68] Residents of Abaco, fearful of the health effects associated with fifty 55-gallon drums of dioxin-based herbicides, threatened to forcefully ship the substances to the Bahamian island of New Providence for the government to handle.

After rejecting the government's plans, reEarth mobilized its staff and contacted activist organizations outside the country in order to leverage support for their efforts. They soon created a transnational activist coalition, which included reEarth, the U.S.-based letter-writing campaign group Global Response, and Greenpeace International. Activists prepared to demand of Owens-Illinois that the obsolete pesticides be "returned to sender" rather than burned or buried in the Bahamas. A good deal of work was required before the campaign could be launched.

From Conflict to Collaboration

From the beginning, the international coalition of environmental justice activists took a diplomatic approach to the pesticide stockpile on Abaco. However, governments and corporations frequently reject such polite overtures by activists, sparking more aggressive responses by grassroots advocates. Surprisingly, this never happened in this case.

In 1991, Greenpeace International's Ann Leonard wrote to Joseph Curry, the vice consul at the Bahamian embassy in Washington, D.C., politely informing his office of the environmental justice coalition's intentions:

> Global Response, an environmental organization based in Colorado, has helped Greenpeace win a number of international environmental victories, including halting shipments of toxic waste from New Jersey to South Africa. This organization has thousands of members throughout the U.S. who agree to write letters to U.S. corporations to convince them to take specific steps to protect the environment. They would like to ask their members to write to the President of . . . the responsible company, and ask him to remove the wastes from Abaco. Such a campaign could result in the clean up of the site at no expense to your government. Of course, Greenpeace will not proceed with any such campaign without consultation with your government.[69]

Leonard's letter to the embassy regarding the campaign made it less likely that anyone in the Bahamian government could later accuse the coalition of being "outside agitators" operating in secrecy, without the state's knowledge. In that same letter, Leonard also thanked Curry for calling her to ask for an update on a bill in the U.S. House of Representatives on banning the international waste trade. What is clear here is that, as is the case in many other instances, global South nations like the Bahamas frequently rely on

Greenpeace for technical information, legal updates, and political influence regarding the waste trade globally and in their own nations.[70] This is one indicator of how well respected Greenpeace International is and how poorly resourced these southern nations are.

Thus, Greenpeace cultivated a mutually beneficial relationship with the state. The Bahamian government, through its embassy in Washington, D.C., maintained regular contact with Greenpeace, as illustrated in a letter addressed to Ann Leonard, regarding the government's commitment to ensuring that the cleanup of the site moved forward as planned: "Enquiries will continue to be made of that department [Department of Environment Health Services] until assurances have been given that the redrumming has in fact been carried out, and the Embassy will continue to forward information on this and other relevant matters as it becomes available."[71]

The diplomacy continued. In March 1992, Global Response's executive director, Malcolm Campbell, contacted Lowell Arbury, a leading businessman on Abaco with the Abaco Wholesale Company. Campbell was seeking advice on how to move forward with a letter-writing campaign:

> Dear Mr. Arbury, Thank you for taking the time to speak with me last evening. . . . What we would need from you would be any background information which would be pertinent to our members' understanding of the issue (certainly given full consideration to not create any negative publicity about the potential hazards on Abaco). We would also need the names of the officials who you would want us to target at Owens-Illinois (and also the Bahamian government who you think would be appropriate). . . . And we would also like for you to tell or indicate to us what you would want our members to ask of the targeted officials in our letters.[72]

In a set of highly respectful moves, these northern TSMOs sought advice from the Bahamian government and a Bahamian businessman to ensure that the process was as collaborative and supported by islanders as much as possible.[73] The activists continued building support among Bahamian leaders. In the spring of 1992, Leonard contacted the director of the Grand Bahama Port Authority, a colleague she had a professional relationship with from an earlier visit to the island over concerns about a proposed waste incinerator, to inquire about how to proceed with the waste removal:

> I have received information from environmental groups in the Bahamas and from the Bahamian embassy here in Washington about an abandoned pile of hazardous wastes and pesticides in Abaco. According to the Ministry of Health, dioxin is present in several of the abandoned herbicides at the site. Some of these pesticides

are banned from use in the United States. Obviously, Greenpeace is very concerned about the environmental and public health threats of these toxic substances. According to recent press reports, the Bahamian government is considering three options for handling the substances: 1) entombing the wastes on site; 2) bringing in a mobile incinerator to burn the wastes on site; 3) transporting the wastes to Grand Bahama for incineration. . . . Since burning or burying this type of waste guarantees both immediate and long-term environmental contamination, Greenpeace would like to see the U.S. company responsible for exporting the toxics to the Bahamas remove all the waste and return it to the U.S. . . . I am very anxious to find out more about the situation so that Greenpeace can work to have the waste removed.[74]

Local activists then made direct diplomatic overtures to the Owens-Illinois Company. When these efforts yielded no response, they zeroed in on the corporation as the offending party and sought to make it the primary target.

A couple of months later, in summer 1992, Global Response and Greenpeace were ready and formally collaborated with reEarth to launch an international letter-writing campaign targeting Owens-Illinois for the cleanup of the pesticides and hazardous wastes. As part of its campaign, Global Response released a "Young Environmentalist's Action" fact sheet about Abaco directed at children. The call to action read, in part: "Please write a polite letter to Joseph Lemieux, the President of Owens-Illinois. Tell him how you feel about protecting the environment. Ask Mr. Lemieux to help the people on Abaco, Bahamas. They are asking him to remove the dangerous pesticides his company brought to their island."[75]

Just twenty-one days after launching the campaign, Global Response received a letter from the Owens-Illinois company stating, "We have decided to take the lead in a plan to properly contain the drums and remove them."[76] This was a total victory in a phenomenally short period of time. The group reEarth rejoiced and gave credit to the company for accepting responsibility and for acting so promptly: "We at reEarth would like to thank Owens-Illinois for recognizing the fact that the immediate priority is the removal of the pesticides."[77] reEarth spokesperson Sam Duncombe wrote a letter of acknowledgment to Global Response's Malcolm Campbell stating, "On behalf of reEarth I would like to thank you from the bottom of my heart for all of your help. . . . Malcolm, to your letter writers, thank them for their efforts and their notes of encouragement—both were appreciated tremendously."[78] In reEarth's newsletter, the group publicly announced the results of this international environmental justice collaboration: "We would like to thank the membership of Global Response for caring about their global neighbors, their Director Malcolm Campbell, and Ann Leonard of Greenpeace for putting us in touch with one another. That's what you call true Global Partnership."[79] Duncombe also publicly thanked Owens-Illinois again for its actions: "Your company has demonstrated true global/corporate leadership by taking the incentive to clean up this situation."[80]

In what might be viewed as an unusual move, Owens-Illinois was also in formal communication with Greenpeace during this process, as evidenced by letters it sent the organization, outlining its plan for environmental cleanup. Owens-Illinois also acknowledged its lack of expertise on non-incineration technologies. As one representative wrote to a Greenpeace staff member, "Regarding the disposition of the herbicides. . . . You mentioned alternatives to incineration for the 2,4,5–T materials [Agent Orange]. I would appreciate if you would send me some information."[81]

In the interest of returning the toxics to the global North, the decision was made to ship the waste to Finland, where it would be incinerated.[82] In January 1995, the *Abaconian* newspaper reported: "The Bahamas government, working jointly with Owens-Illinois, has had all the hazardous waste disposed of from the sugar mill site. The firm Environmental Service of America had the contract to remove it at the cost of $370,000. The company opened all the old drums and repackaged the most hazardous wastes in special drums which were shipped to Finland for disposal. Other wastes which were not toxic were disposed of in the Bahamas."[83]

The Bahamian prime minister remarked at the time, "I consider the clean-up of the sugar mill site a major health initiative. . . . This island has been threatened with a potential health crisis, too serious to contemplate."[84] The government and Owens-Illinois agreed to an even split of the cleanup costs.

Conclusions

What is significant about this case is that the government appeared to cooperate fully with activists to remediate the waste. It is perhaps unusual that activists were in such close and cooperative contact with the government in the Bahamas and the Bahamian embassy in Washington, D.C. What is no less surprising is that activists had a similar relationship with the Owens-Illinois Company in Toledo, Ohio, and with Bahamian businesses like Abaco Wholesale. Very likely, this was in the interest of promoting tourism and preserving the Bahamas' image as a safe site for foreigners to visit. For example, the local Bahamian partner in the environmental justice coalition, reEarth, publicly supported the government's plans to strengthen the Bahamas' efforts at remaining a major site for ecotourism. In their newsletter, published during the time of the Owens-Illinois pesticide case in Abaco, reEarth stated, "Congratulations are in order to our brand new 'Sunshine' Prime Minister—The Honorable Hubert Igraham. Our new Prime Minister has pledged to make the Bahamas the Eco Tourist destination in this part of the world. All Bahamians young and old have the power in themselves to make that dream a reality. It is up to each individual to make that fundamental change and to promote our beautiful country and to be proud leaders of Eco-Tourism."[85]

Rather than assuming a progressive position on the part of the Bahamian government, a shrewder analysis might consider this environmental cleanup as damage control in a highly image-conscious nation dependent on its sustained appearance not only as a tourist destination but as an ecotourism destination. That is, the Bahamian government was quite aware of the ways in which foreign investors and tourists imagine and construct the natural beauty of the Bahamas and of the material consequences. For Owens-Illinois's part, the company should be acknowledged for its efforts to remove the wastes from the site. A combination of the international environmental justice coalition putting public pressure on the company, the Basel Convention's expectation that exporters of hazardous wastes would be responsible for their cleanup, and Owens-Illinois's plans to reinvent itself entirely as a packaging company (which meant, in part, it would want to settle any outstanding liabilities such as this one) converged to push the company toward corporate environmental responsibility. Pushing the responsibility for the waste cleanup on the company, the environmental justice coalition amassed support locally and internationally to target Owens-Illinois, and it won. Without the activist coalition's international mobilization, this case would certainly not have been addressed so soon, if at all. Activists and journalists also invoked the double standard involved when the United States banned dioxin, while a U.S.-based corporation was allowed to maintain these same wastes on an island nation populated by black and poor people—a clear recognition of environmental injustice that would make for bad publicity for the company and the governments involved.

The case of Owens-Illinois also underscores the problem of obsolete pesticides in global South nations, particularly in the Caribbean. During the past several decades, nearly all forms of agrochemicals and pesticides have been applied to agricultural lands throughout the Caribbean region. This is particularly troubling because the agricultural sector has historically been at the core of the region's economies, and that sector has come to depend on imported pesticides.[86] Although many Caribbean nations now have pesticide control regulations, enforcement is virtually nonexistent, as there are few institutional and economic incentives for doing so.[87]

Stockpiles of obsolete pesticides are a major problem throughout the South. Pesticides become hazardous wastes through a number of means, including excessive or unsolicited donations from northern nations and aid agencies, inadequate storage facilities that allow for deterioration of the chemicals and corrosion of containers—which results in leaking into ground water sources and the air—or storage beyond the usual expiration date of two years. And although Caribbean nations have shouldered a major share of this problem, no region of the globe has more pesticide wastes stockpiled than sub-Saharan Africa.

Africa and Pesticides

Every nation in Africa has been documented as a site hosting obsolete pesticide waste stockpiles.[88] These are sites where pesticide drums are stored for at least two years, in eroding containers, allowing the toxins to leak into the surrounding ecosystems. To address this massive problem, a group of international NGOs, governments, industries, and IFIs partnered to raise funds for the cleanup of all obsolete pesticide stocks in Africa. The group came together as the Africa Stockpiles Programme (ASP) and planned on completing the cleanup the wastes in an environmentally sound manner within fifteen years of its founding. The ASP was started by the World Wide Fund for Nature Toxics Team and Pesticide Action Network (PAN) UK. Realizing that other major stakeholders in the development sector would need to join the effort, they invited the United Nations Food and

Agriculture Organization, United Nations Environment Programme, United Nations Industrial Development Organization, and the Economic Commission for Africa to the table. Several other nations, international agencies, and IFIs soon joined the effort.[89] Experts estimate the volume of obsolete pesticides in Africa at more than 50,000 tons and the cleanup effort is expected to require US $175 million.

Since most African nations do not possess the technology or expertise to dispose safely of these pesticide stocks, the ASP agreed that the wastes should be sent North to more industrialized nations. There were early precedents for this program, in addition to the case in Abaco, Bahamas. For example, in 1991, the United States, Germany, and the Shell Corporation—the manufacturer of the persistent organic pollutant dieldrin—collaborated to remove fifty-four metric tons of that pesticide from the African nation of Niger. In an officially endorsed return-to-sender process, the waste was returned to Holland, the nation of its origin. Many heads of state across Africa have stated their preferences for return-to-sender solutions to the pesticide crisis on that continent.[90] For example, Ekokem, a Finnish firm, was contracted to conduct a cleanup of pesticide wastes in Ethiopia, a nation that the United Nations Food and Agriculture Organization says is the hardest hit of all countries in Africa. Ekokem will return the wastes to Finland for incineration.[91]

In 2003, CropLife International, a global federation representing agro firms, pledged $30 million toward the ASP, which will allow the process to move forward more rapidly than previously expected.[92] CropLife is led by companies such as Monsanto, Bayer CropScience, and Dow Agro Sciences, all of which have come under intense criticism for selling highly toxic pesticides in the global South. It could be argued, then, that the CropLife pledge represents some form of compensation for the role these companies played in producing pesticide stockpiles. The companies agreed to contribute to the effort in order to remediate the particular pesticides they manufactured, and the chemicals will be incinerated in a northern nation. High-temperature incineration in especially designed hazardous waste incinerators is the currently recommended method for disposal of obsolete pesticides, as outlined in United Nations Disposal Guidelines issued jointly by the Food and Agriculture Organization, the UN Environment Programme, and the World Health Organization.

While the return-to-sender approach is laudable and clearly represents the successful institutionalization of activist tactics, incineration of these pesticides is unacceptable to many activists at the International POPs Elimination Network and GAIA, who oppose the ASP for that specific reason. As one activist told me, "Needless to say, that sticking point has created a lot of tension among international environmental organizations, but we just cannot support incineration."[93] The World Bank has, for many years, supported and funded incineration projects in Africa, Asia, and, increasingly, Central and Eastern Europe—all in poor nations. According to a 2001 survey by the consumer advocate organization Essential Action, at least seventy-nine World Bank projects in forty-seven countries involved incineration in recent years.[94] This is problematic not only because it is environmentally unsustainable and unjust for the residents of these nations to shoulder this toxic burden, but also because where the World Bank provides funding for chemical waste incinerators and cement kiln burners, one often finds pesticides the bank has also helped bring into those nations. For example, between 1988 and 1995, the World Bank provided funds for the purchase of $250.75 million worth of pesticides, mainly manufactured by TNCs in the global North, which were transferred to southern nations.[95] Thus, the World Bank creates a situation of double jeopardy, wherein regions and nations where pesticides are being exported are also witnessing a rise in major incineration projects.

Despite the ASP's work, there remain attempts to handle abandoned pesticides in Africa through less than sustainable means and through processes that would compound existing environmental injustices. Recently Mozambique was the site of a major conflict concerning such a case.

Something Toxic from Denmark: Mozambique's Battle with Foreign Pesticides

Mozambique is a nation of 19 million people, 70 percent of whom live off the land. Located in southeastern Africa, it is the world's ninth poorest nation. Mozambique is slowly rebuilding itself after five centuries of brutal colonization by Portugal, followed by seventeen years of civil war that resulted in the deaths of 1 million persons. Former independence fighters with the group FRELIMO won the country's first democratic elections in 1994, and UN peacekeeping forces finally departed one year later. Since that time, Mozambique has enjoyed relative peace. Even so, the average Mozambican's life expectancy is just forty years, and the citizenry experience grinding poverty on a daily basis.[96] The U.S. ecological footprint is 23.7 acres per capita; a sustainable footprint in that nation would be 4.6 acres. Mozambique represents the other end of the scale, with an ecological footprint of 1.3 acres per capita. Unfortunately, the reason for this lighter footprint is that there is so much poverty and so little industrialization occurring in Mozambique.[97] Despite this harsh reality, civil society organizations are beginning to emerge and thrive in this once-chaotic place.[98] And the first signs of civil society growth in Mozambique sprang forth from an international struggle for environmental justice.[99]

A Toxic Discovery, and a Community Mobilizes In 1998, in the capital city of Maputo, community activist Janice Lemos read a story in *Metical*, an independent local newspaper, about an effort funded by a Danish international development agency (Danida) to incinerate nine hundred tons of obsolete toxic fertilizer and pesticide stocks in a cement factory in the southern city of Matola. Matola is a suburb fifteen kilometers outside Maputo. Danida sought to donate a hazardous waste incineration facility that would be housed in the cement factory, which the aid agency would also pay to have retrofitted for the operation. Lemos wrote to the newspaper for more information, but none was available. She then contacted Greenpeace International headquarters in the Netherlands, where someone informed her that two U.S.-based toxic waste activists would soon be visiting South Africa, and they might be able to travel to Maputo and Matola if Mozambican community leaders would invite them. With the help of Greenpeace and Oxfam Community Aid Abroad, Lemos and fellow concerned residents met with U.S. activists Ann Leonard (then with the group Essential Action) and Paul Connett (a St. Lawrence University chemistry professor and renowned expert on and opponent of incineration) and Bobby Peek, a South African toxics expert and activist (with Environmental Justice Networking Forum, EJNF).

The visiting activists were concerned because they had documentation that cement kiln incinerators produce a range of deadly toxins such as dioxins and furans. In fact, scientists estimate that 23 percent of the world's newly created dioxin comes from cement kiln incinerators alone.[100] Prior to their arrival, the visiting activists were able to access documents about Danida's plans and had additional information about the proposed project. Peek stated, "Whether or not anybody actually became concerned about the issue . . . we strongly felt that we had the moral obligation to pass on what we knew about the plan, and the real risks of cement kiln incineration. They had the right to know. As we feared, almost nobody had heard about the project at all."[101] This lack of public knowledge was particularly disturbing because Danida has a policy of "actively involving individuals, non-governmental organizations and associations and businesses formally and informally in formulating and implementing environmental policies."[102] Yet few people in Maputo or Matola had heard anything about the project from Danida. In fact, the foreign visitors were the only people at the meeting who had seen or possessed a copy of the short environmental impact assessment that Danida had prepared. The report was written in English, although the official language of Mozambique is Portuguese.[103] One local activist remembered, "Only a few of us could manage to read the report and . . . do a brief analysis."[104] Connett denounced the entire project: "In the United States or Canada, those proposing a new toxic waste facility would be obliged to fully discuss all of the alternatives, all of the risks, and would have been required to hold several public hearings before decisions could be made about a particular disposal method. The environmental assessment and public involvement in this project is a sham."[105] Danida's environmental impact statement claimed that no serious environmental impacts would result from the incineration of the pesticides.[106]

The visiting activists also informed the Mozambican citizens of the questionable record of Waste-Tech Ltd., a South African firm that was to be contracted for the Danida effort. At that time, Waste-Tech was seeking to import foreign waste into South Africa, a clear violation of law there, and was the subject of an investigation by the South Africa Human Rights Commission concerning possible human rights abuses in the case of two incinerators it had located within a short distance of a poor community. The company was also dealing with other legal investigations being conducted by the South African Department of Water Affairs and Forestry.[107]

Mauricio Sulila, one of the local community leaders from Maputo present at the meeting, later told a reporter "When we explained [to others attending the gathering] that the government had decided the factory would burn toxic waste, they became terrified."[108] Local people had already suspected that there were toxic materials at the site because, as Sulila recalled, earlier flooding in the area had prompted residents to pump the water into a nearby swamp where, soon afterward, "someone ate a fish caught in this swamp and died."[109]

At the meeting with U.S. and South African activists, local residents and community leaders decided to create an organization to address the problem of environmental hazards in the area. Mazul, one of the attendees, who was also an artist, explained that since the citizens had been kept in the dark by the Danish and Mozambican authorities at the Environment Ministry, the group should be named *Livaningo,* which translates to "all that sheds light" in Shangaan, one of many languages spoken in that region of the country. Mauricio Sulila was appointed the group's general secretary. Janice Lemos and her sister Anabela joined the group's leadership as well.

Action and Networking at Multiple Scales Soon after, Livaningo grew in its influence on the local and national government and took on the case of the pesticides in Matola with great determination. They organized public gatherings and meetings, brought their concerns to local residents and businesses, and made strategic use of the independent press. They held some of the first public demonstrations in postrevolution Mozambique. Sulila explained, "It is important to say that Livaningo was the first organization in Mozambique to really challenge the government."[110] Livaningo was eventually able to get another firm to conduct an independent environmental assessment of the project. Anabela Lemos proudly recalled that "their conclusion was completely what we thought from the beginning: under no condition should the cement factory should be turned into an incinerator."[111]

However, the government refused to consider the independent environmental assessment. Livaningo then tried to get an audience with the Danish embassy in Maputo but was refused. In fall 1998, members of the Danida board of directors visited Maputo but rejected Livaningo's request for a meeting. In response, Livaningo decided to elevate the struggle and go to the source. Aurelio Gomes of Livaningo and Bobby Peek of EJNF in South Africa traveled to Denmark to address the Danish Parliament about Danida's pesticide incineration project. As Gomes stated on arrival in Denmark with regard to Danida's earlier refusal to meet with them, "This won't prevent us from voicing our concerns, therefore we've come to Copenhagen today to provide the Danish government with information to justify the immediate halt and rethinking [of the incineration effort]."[112] Although the Parliament granted them an audience, the legislators made no effort to intervene. Even so, this was a significant moment in the formation of the international environmental justice collaboration, because Mozambican, U.S., South African, and Danish activists were working together in tight coordination. International allies such as Greenpeace and the Joint Oxfam Advocacy Program (JOAP, Mozambique) donated the funding support for these activities. Sulila remembered:

> That was great. . . . After that, the Mozambique government opened up the door a little. We explained to them that we will not give up, we will not be intimidated. We continued to make pressure, to make noise, to hold international meetings and meetings at the local level. We were working with several organizations, especially Greenpeace Denmark. JOAP's support was fantastic. Say we need to do a demonstration in two days, they were able to provide funds to advertise in the newspaper. When we needed to travel to Denmark, JOAP funded us. It's not a lot of money, but it is at the right time, when we really need it.[113]

When asked later how Livaningo organized so effectively on an international scale, Anabela Lemos stated, "It is mostly through the Internet. But whenever we campaign, we make some noise here in Mozambique, and at the same time we have the international network. When our government told us to stop complaining, we went to Denmark and we spoke to the people *there,* and we realized that, as a result, they started to listen to us *here.* So we realized then that we couldn't just do a campaign here, but instead we had to work both ways, here in Mozambique and in Denmark."[114]

Activists with Greenpeace Denmark were critical to the campaign as well. While the Danish government initially refused accountability for the pesticides, Danish activists took responsibility for their nation's complicitness in this act of international environmental injustice. Greenpeace Denmark staff member Jacob Hartmann commented on the hypocrisy involved in his government's strong advocacy for the Basel Convention on Transboundary Hazardous Wastes while also encouraging the incineration of pesticides in Matola: "Considering that Denmark is one of the countries that have taken the lead on this vital treaty, it makes little sense for Denmark to advocate for an elimination of POPs globally while promoting new sources of the worst of them in Mozambique."[115]

Denmark's development minister, Poul Nielsen, denied that his country was seeking to impose incinerators on Mozambique,[116] but activists found this statement suspect, given that Danida had funded a failed incinerator in India in 1986, and because Denmark was considering support of garbage incinerators in Zimbabwe and Tanzania in 1998, the year the conflict in Mozambique ignited.[117]

Political Economic Opportunity Structures and Environmental Injustice Activists in the coalition consistently called for the pesticide incineration project to be halted, for the pesticides to be exported to a global North nation, for the wastes to be disposed of using nontoxic, nonincineration technology, and for all the costs to be borne by the companies that produced the chemicals in the first place.[118] Articulating a political economic opportunity structural approach, they sought to make both corporations and states responsible for this problem. And although Livaningo activists were only beginning civil society organizing on environmental justice issues, they were familiar with environmental injustice, since this was something that has been widespread in the region. Anabela Lemos remarked, "In South Africa, always the dumping sites are near the poor people. And we have a waste dump here in the city and there is a concentration of poor people there, so it's the same thing here. The poor, they always get the waste."[119] As to the origin of the pesticide wastes and the process that led to this particular case of environmental injustice, Lemos argues: "The pesticides came from a lot of different countries. A problem for a country like ours is that there's no control over donors, what they give and not give, there are no controls imposed. So if there's an agricultural project, for example, and they need pesticides, then they have to receive what they give to them. There's no control. . . . And, from my way of seeing, this is a result of over production in a first world country. They don't know what to do and they just dump here."[120] The lack of regulation over the trade in agrochemicals must be addressed if long-term environmental justice is to be attained.

Resolution After two years of campaigning locally in Matola and with the national government in Maputo, Livaningo had its first major breakthrough. The Mozambican government agreed to a return-to-sender arrangement and allowed the chemicals to be shipped to a global North nation, the

Netherlands, for processing and disposal by hazardous waste treatment firms there.[121] While the Mozambican government did have to pay some of the costs, Denmark shared the expenses. And despite environmental justice activists' hopes that nonincineration technologies would be used, some of the wastes were indeed incinerated in Europe. Even so, the environmental justice coalition achieved its primary goal of return to sender: exporting the wastes to a global North nation. The international environmental justice movement had once again succeeded at having its approach accepted by states and corporations implicated in transnational environmental injustices. Livaningo reached out to a broad group of established NGOs, including the Environmental Justice Networking Forum (South Africa), Essential Action (United States), Greenpeace International (Netherlands, Denmark, and Brazil), the Basel Action Network (United States), and Oxfam's JOAP (United Kingdom), to amplify its voice and augment any leverage it already had to achieve one of the most impressive global-local environmental justice collaborations in the movement's history. The South African–based EJNF lent a critical African presence to the struggle. No less important was the legitimacy that Livaningo provided for its international partner organizations and activists who might otherwise be viewed as "outside agitators" in Mozambique. And Greenpeace Denmark provided much-needed credibility for Mozambican activists confronting the Danish government. Drawing on local, regional, and international activist support, as well as international law and aggressive movement tactics, the coalition succeeded. These external resources were critical to the campaign, but the local activists' level of determination and commitment to the struggle were what ultimately sustained the effort. Livaningo's Aurelio Gomes remarked, "We have nothing against Denmark, and hope they have nothing against us. We just want them to understand that here in Mozambique, while we may not be wealthy, we will never compromise our health—that is all some of us have."[122] And Livaningo activist Anabela Lemos commented, "We just decided that we would not fail, although there were many times when it looked as if all hope was lost."[123]

Mozambique is a young democracy. Its government is still slowly becoming accustomed to the idea of being challenged by civil society groups, whether inside or outside its borders. As Anabela Lemos commented, "Mozambique is a country where people are scared to speak out, and still today, but it is getting better. We are going through democracy after so many years. We are the only NGO doing this work. If something is wrong, we speak up, we don't talk just for talking's sake. When we speak, we know we are right and we know we have to say it."[124]

Next Steps for Mozambique

The effort to halt the incineration project and export the pesticide wastes from Matolo, Mozambique, was successful and was a pleasant surprise for people throughout the international NGO community as they witnessed activists from one of the world's poorest nations exert uncommon political leverage.

Since this unprecedented success, Livaningo has used the opportunity to broaden its focus beyond toxics to other environmental justice struggles in the region. It is pursuing projects aimed at introducing ecologically sound waste management systems in health care institutions in Mozambique, in collaboration with international activist groups like Health Care Without Harm.[125] Its members are also working to oppose harmful development projects like the Mepanda Uncua Hydroelectric dam on the Zambezi River in Mozambique, in partnership with global advocacy NGOs like the International Rivers Network.[126] Livaningo is also combating oil extraction efforts by TNCs in southern Africa, which would pollute the air, land, and water and return few economic benefits to the people of the region.

Livaningo's victory in the pesticide incinerator case is credited with opening a broader political space for other civil society groups and social movements to work in Mozambique on a host of social concerns. Organizations working on HIV/AIDS, human rights, land rights, and global economic justice efforts now enjoy greater support and freedoms as a result of the political space Livaningo opened. In other words, they forced access into the nation's political economic opportunity structure by challenging and transforming the structure itself. As Anabela Lemos stated, "It is true that we opened things up for people in our nation because we are not scared to speak out and to raise our issues. We think we have the right to do so. And I think that civil society has to get involved, we can't just sit on our hands and complain. If something is wrong, we should work for it. I think people should start to realize that to have big changes we have to give a lot up and we have to sacrifice. . . . You should not be scared. If you are right, then you have the right to speak, and I think it does make a difference."[127]

The Bahamas and Mozambique

Comparing the Bahamas/Owens-Illinois case with the Mozambique/Danida case, some instructive differences and similarities become clear. In the Bahamas case, with the exception of Ann Leonard's visit years before this case emerged, activists from the global North did not make personal visits to the islands to make face-to-face contact with their activist partners there. Instead, through telephone calls and letters between U.S. and Bahamian activists, to local business leaders, and the Bahamian government (through the embassy and the port authority), the activist coalition gathered enough information and support to launch a global letter-writing campaign directed at Owens-Illinois. Although the same principles of collaboration applied in both the Bahamian and Mozambican cases, the former struggle was comparatively less complex, for four reasons. First, there was a single corporation and government made responsible for this environmental injustice. Second, there was a moderately strong local civil society and activist infrastructure in place in the Bahamas. Third, the corporation was willing and able to perform the remediation and export of the wastes. And fourth, the state had a clear interest in maintaining its image as a tourist's paradise, as tourism is responsible for 60 percent of the Bahamas' gross domestic product and employs at least half of the nation's labor force.[128]

The case of Livaningo was much more complicated for a number of reasons. First, there was little to no civil society infrastructure in Mozambique at the time, and the democratic political system there, only four years old, was unaccustomed to handling demands from community constituents. Second, the citizenry of Mozambique were not only relatively inactive politically, they suffered under far greater, deeper poverty and international isolation than the citizenry of the Bahamas. Third, while many corporations were responsible for manufacturing and exporting the pesticides in the Maputo case, no one had any solid information connecting the waste to particular firms, and none stepped forward the way Owens-Illinois did. This also meant that since the activist targets were a Danish national development agency and the government of Mozambique, neither party felt entirely responsible for removing the wastes. Hence, they sought third-party nations and corporations willing and capable of storing and treating the pesticides. Fourth, because both the Danish and Mozambican governments were initially unsympathetic to activist demands, this case required activists from the North to travel to Mozambique and required Mozambican activists to travel to Denmark; external pressure was necessary. Danish environmentalists had to pressure their own government as well. Considering this matrix of barriers, it is that much more remarkable that the coalition achieved its goal in just two years.

Both cases have much in common as well, which speaks to the three arguments I am advancing in this book. First, we consider racism and class inequalities. reEarth and Livaningo were both struggling against international environmental inequality, where excessive volumes of highly toxic pesticides were exported by northern corporations, states, and aid agencies to nations populated largely by poor, non-European peoples. Rather than being color blind, states, aid agencies, and corporations have pushed poisons on these nations because it is both possible and profitable to do so, while increasingly difficult to do so elsewhere. Globally there is a strong correlation between poverty and the presence of non-European peoples,[129] so while these policies may not be explicitly racist, the results are. And since the communities involved in both cases were poor, class exploitation was at work as well. Race, class, and nation are thus intimately linked through the toxicity of social domination.

We should also consider ecological modernization versus ecological decline. The contention that ecological conditions in the world are improving is both supported and contradicted by the reality of environmental inequality. That is, the thesis seems plausible when we observe environmental improvements taking hold within a specifically global North geographic context. However, ecological modernization falls short because such improvements in the North are largely due to the actions of institutions that are shifting much of the most toxic hazards southward.[130] One's view of ecological modernization therefore depends on what spatial scale one considers and whose environment one is concerned about. Both Mozambique and the Bahamas suffered in these cases from international environmental inequality. In both struggles, political authorities made clear that the preferred solution was incineration of the pesticides. As noted earlier, the waste incineration process increases ecological risk by creating additional and more highly toxic substances. This course of action would compound an already hazardous situation. As Livaningo's Mauricio Sulila remarked, "It would indeed be ironic if . . . Africa [were] saddled with a new legacy of polluting incinerators and cement kilns set up to destroy an earlier legacy caused by pesticides dumped on Africa by aid and trade organizations."[131] That is, in the absence of a strong international environmental justice movement, political and economic authorities would likely resort to "solving" one environmental injustice by imposing another. This is an example of how systems of inequality intersect, placing layers of oppression on each other even as solutions are sought.

Finally, we must consider the political economy of environmental justice struggles. In both cases considered here, states and corporate actors were responsible for the transfer of toxics from North to South, and these same stakeholders were critical to the resolution. Agrochemical corporations shape the ecosystems of southern nations when toxics are exported and used in agriculture. And as much as they may assist farmers with production, pesticides also cause great harm to the land, water, air, and human health. Corporations manufacture these chemicals that frequently take on the physical reality of environmental inequality. It is the nexus of state and transnational corporate power that makes environmental justice struggles a challenge for local activists, who may have little or no way to influence states or institutions beyond their borders, and often have minimal power to do so within their borders. By collaborating to build transnational movement networks, activists can shame and persuade states and corporations responsible for the transfer of toxics across borders to undo the damage they created. When activists recognize that states need corporations for private investment and revenue generation and that corporations need states for infrastructure development and security, points of access in this political economic opportunity structure open up.

Local and Global Resistance to Corporate Domination and Poisoning

The transnational environmental justice movement networks examined in this chapter advocated a return-to-sender approach for toxic wastes dumped or abandoned in global South nations. The hope is that as more such campaigns succeed, "the cost and visibility of storage in the North acts as an incentive for waste producers to adopt clean technologies, an approach which would have prevented such problems in the first place."[132] In the meantime, transnational environmental justice movement networks are focusing on the corporations and their partner states that push pesticides on poor people and people of color around the world, thus confronting an unequal global political economy.

Targeting the Global Political Economic Opportunity Structure

In 2003, activists from across Asia participated in "Poisoned and Silenced: A Public Hearing of the Impact of Pesticides on Human and Environmental Health," held at the Asian Social Forum in Hyderabad, India. At that event, workers exposed to pesticide poisoning testified before an audience and a panel of "judges" about their illnesses and the toll these chemicals had taken on their lives. At the conclusion of the testimonies, the judges declared a unanimous verdict of "guilty!" They issued a statement that "the companies responsible for pesticide poisoning, and the governments that are their accomplices, are liable for violations of the people's right to safe and healthy working conditions and environment." Kaveri Dutt, a human rights lawyer who participated in the session stated, "Today is not about being in a court of law. It is a Court of Conscience. This is more important than anything else. The pesticide workers must come together and tell the companies, especially Syngenta, that they have committed murder and will be sentenced. If pesticides have diseased us, it has diseased them more with an incurable disease: that of greed."[133] The political economic opportunity structure critique is alive and well among environmental justice activist networks in the South.

In 2002, the government of Malaysia decided to ban the pesticide paraquat after more than forty years of use in that nation. Paraquat is the world's second most widely used herbicide, after glyphosate (known by its trade name Round Up) and is viewed as highly toxic and hazardous. This was a momentous occasion because it was the result of years of organizing and campaigning by environmental justice, human rights, and workers' rights activists across Southeast Asia. It was also a key victory for activists because they had targeted the Syngenta Corporation, the largest producer of paraquat, for many years.[134] In response to this decision, Irene Fernandez, activist and coordinator of Tenaganita, a workers' rights NGO in Malaysia, stated, "We congratulate the Malaysian government for its decision, and its consideration to protect workers' health. We want all the manufacturers, especially Syngenta, to respect the government's decision by ceasing production of paraquat, and we demand that they recall all stocks of paraquat immediately! While this is a great step for the country, paraquat is still a problem elsewhere. We will continue the global campaign on paraquat, and its biggest manufacturer, Syngenta."[135]

Later that year, the Berne Declaration, a Swiss public interest NGO, joined with Pesticide Action Network Asia and the Pacific, Pesticide Action Network United Kingdom, the Swedish Society for Nature Conservation, and Foro Emaus (a Costa Rican NGO) to launch a campaign against paraquat and Syngenta, beginning with its report, "Paraquat: Syngenta's Controversial Herbicide." The coalition started an e-mail campaign to target the company, in which they focused on Heinz Imhof, the board chair, to stop production of paraquat altogether. Not surprisingly, the industry fought back. Mageswari Sangaralingam, a research officer for the Consumers' Association of Penang (CAP, Malaysia) noted: "Since the time of the ban, lobbying by the pesticide industries and allies undermined the decision made by the Malaysian government. Considering the serious health and ecological concerns for paraquat effects, CAP strongly recommended to the Malaysian government that it maintains its ban on paraquat. We view that Malaysian farmers should be educated as to the dangers of synthetic chemicals and urged to practice sustainable agriculture using biological pest control or environmentally benign herbal pesticides."[136]

The industry lobby demonstrated its power in October 2003 when a European Union standing committee voted to allow paraquat to be re-registered in the EU after it had been banned years before. This decision was controversial, and many EU countries and environmental NGOs protested loudly, since paraquat is banned in Austria, Denmark, Finland, and Sweden and restricted in Germany.[137] Environmental inequality reveals that politically and economically vulnerable populations are frequently the miner's canary[138] in that when we observe danger and risks affecting these populations, the rest of the world may soon follow. Similar to the risk society thesis and the circle of poison model of pesticide migration, these risks can always serve as a threat to the world's dominant communities, which is what happened in the case of paraquat. In this case, the poisons migrated back to Europe through both natural ecosystem functions and as a result of political economic policymaking that formally allowed their use in the EU once again.

Pesticides as Political-Economic Violence

Pesticides are used to destroy life. As Arjunan Ramasamy, a plantation worker in Malaysia so aptly told the Syngenta Corporation's shareholders at one of their recent annual meetings, "A poison is a poison. It is made to kill."[139] Thus, perhaps no one should be surprised that many antipesticide activists are also deeply opposed to militarism and state violence. These activists are keenly aware that many chemicals (and the companies that produce them) used in warfare are similar to those used in chemical-intensive agriculture. Weeks after the United States embarked on its war in Iraq in 2003, the Pesticide Action Network Asia and the Pacific issued a press release condemning that action, because it is well known that the U.S. military uses toxic materials in warfare. Rejecting the U.S. government's argument that they were searching for weapons of mass destruction, the network accused the Bush administration of using "poisons of mass destruction" to wage war. The statement challenged the war as a "crime against humanity. It is an unjust, immoral and illegitimate war that is causing destruction of lives, property and the environment."[140]

Pushing for Solutions

Activists working to ban and remove pesticides from the global South are part of a broader effort in which advocates are pressing for sustainable agriculture and food production methods around the world. One technique that many activists and policymakers support is integrated pest management (IPM). IPM is an ecologically and fiscally low-impact method of controlling agricultural pests that emphasizes local knowledge and the empowerment of workers and farmers. The goal is to enable farmers to maintain their crops with minimal use of chemical pesticides. Not only do many grassroots environmental justice activists support IPM, the United Nations Food and Agriculture Organization also promotes it in more than forty nations. The United Nations also invokes IPM in its Agenda 21, in which the proposed direction for sustainable agriculture is based on the principles of IPM.[141] This level of acceptance of IPM is evidence that the global environmental justice movement has achieved important successes in policy making and programming among global multilateral institutions, in addition to doing the work of repelling and returning pesticides and stockpiles from South to North through grassroots and transnational advocacy campaigns.

6

Electronic Waste: The "Clean Industry" Exports Its Trash

Modernization today is almost universally equated with the degree to which a nation is integrated into the global economy. One of the key paths to this integration is through a commitment to technology and infrastructure associated with the new economy. Modernization and information technology are now considered symbiotic.[1] The digital divide between those populations with or without access to this technology has become a public policy concern. Global North nations are heavily wired, while access to this technology in most southern nations is considerably lower, although growing.[2] In the North, transportation, commercial, educational, health care, communications, military, media, and governmental infrastructures rely on computerization for data management, processing, and storage. The various sectors associated with high technology form the largest manufacturing industry in the world.[3] This industry is also one of the most rapidly changing in human history. For example, the speed of the microchip—the brain in most computerized devices—has doubled every eighteen months for more than three decades.[4] This also means that consumer electronics, particularly personal computers, become obsolete quite rapidly.

When computers and other electronics goods are discarded, this "e-waste" is often shipped to urban areas and rural villages across Asia, Africa, and Latin America, where residents and workers disassemble them for sale in new manufacturing processes or where they are simply dumped as waste. Because each computer monitor (the cathode ray tube) contains several pounds of highly toxic materials, this practice creates a massive transfer of hazardous waste products from North to South and is responsible for harming public health and the integrity of watersheds in Bangladesh, China,

India, Nigeria, Pakistan, the Philippines, and Taiwan, for example. This process has also transformed rural or semirural regions into emerging urban spaces and created greater population and pollution densities in already heavily urbanized locations. The problem is expected to worsen because e-waste is the fastest-growing waste stream in industrialized nations.

In this chapter I examine the origins of this problem and how the low-tech side of high tech harms public and environmental health in spaces around the globe where the poor, the working class, and people of color are concentrated. This form of toxic dumping reveals that global environmental inequality is not easily dismissed as the result of older technologies being rejected and sent South; rather, these toxics are part of the technological cutting edge that defines the industrial and consumer vanguard of the new global economy. And if this economic sector is engaged in exploitative practices that are so deeply unequal across race, class, and nation, this should give pause to scholars and policymakers who would hope that modernization would produce both greater environmental efficiencies and social equality.

A sophisticated transnational effort has come together to document these problems, and activists have had success at changing corporate environmental policies and passing local, national, and international legislation to address the worst dimensions of the e-waste crisis. I consider these dynamics in the context of social movement theory and grassroots resistance methods in a global political economy. I argue that local and global environmental crises are best addressed through a combination of strategies that target both states and large private corporations. I also argue that the political economic opportunity structures that activists confront have strong negative impacts on racialized and poor communities in the South.

The Problem: Electronic Waste

E-waste encompasses a broad and growing number of electronic devices ranging from large household appliances such as refrigerators, washers and dryers, and air conditioners to cellular phones, fluorescent lamp bulbs, and personal stereos. At one time, consumers bought a television or stereo and expected it to work for at least a decade, but today the intense pace of technological evolution reduces all things to the status of disposable goods.[5] People no longer take their broken toaster, VCR, or telephone to a repair shop. Replacing these products is often more convenient and more cost-effective. These electronic gadgets may offer greater convenience as they get smaller, cheaper, and faster each year, but they leave a legacy of environmental harm.[6] The most visible and harmful component of e-waste today is the personal computer.

European studies estimate that the volume of e-waste is increasing by 3 to 5 percent per year, which is almost three times faster than the municipal waste stream is growing generally.[7] The United States remains the chief source of e-waste, since U.S. consumers purchase more computers than the citizenry of any other nation. The situation in Europe is not much better. For example, an estimated 1 million tons of electrical waste is produced in the United Kingdom each year and is increasing by 5 percent annually. Of this volume, 90 percent is thrown into landfills or incinerators.[8]

The e-waste problem is a crisis not only of quantity but also a problem of toxic substances: the lead, beryllium, mercury, cadmium, polyvinyl chloride plastics (PVCs), hexavalent chromium, and brominated-flame retardants (BFRs) pose major occupational and environmental health threats.[9] Computer or television displays contain an average of four to eight pounds of lead each. The 315 million computers that became obsolete between 1997 and 2004 contained more than 1.2 billion pounds of lead. When these components are illegally disposed of and crushed in landfills, the lead is released into the environment, posing a hazardous threat to current and future generations. Consumer electronics already constitute 40 percent of lead found in landfills. About 70 percent of the heavy metals (including mercury and cadmium) in landfills come from electronic equipment discards. These heavy metals and other hazardous substances can contaminate groundwater and pose other environmental and public health risks. Lead can damage the central and peripheral nervous systems, the blood system, and kidneys in humans. It also accumulates in the environment and has highly acute and chronic toxic effects on plants, animals, and microorganisms. Children suffer developmental effects and loss of mental ability even at low levels of lead exposure. Mercury leaches when certain electronic devices such as circuit breakers are destroyed. The presence of halogenated hydrocarbons in computer plastics may result in the formation of dioxin if the plastic is burned. The presence of these chemicals also makes computer recycling particularly hazardous to workers and the environment.[10]

Consumers are at the center of the problem as well. The throwaway consumer culture in the global North persists and is encouraged by both government and industry. Since the first commercial cell phone services were introduced in the United States in 1983, the use of cellular phones jumped from 340,000 in 1985 to more than 203 million in 2006.[11] Although small in size by comparison to computers, cell phones comprise an increasing percentage of the electronics—and general—waste stream in the United States and around the globe. Recently millions of cell phone users applauded the Federal Communications Commission ruling that consumers can keep their phone numbers even when they switch cell phone companies. Unfortunately, this policy will also increase the level of e-waste associated with these devices, since many cell phone users will have to purchase new phones when they switch providers. The U.S. Environmental Protection Agency estimated that 130 million cell phones were discarded in 2005, resulting in 65,000 tons of e-waste. There are a number of cell phone recycling operations up and running, but "they are no match for the growing heaps of cell phone trash," according to Inform, a New York–based environmental research organization. The USEPA says cell phones present an environmental hazard because they contain lead and BFRs, mercury, cadmium, and gallium arsenide.[12]

In chapter 1, I discussed the relative merits of the treadmill of production and ecological modernization theories. The trends in electronics production, consumption, and disposal are revealing in that regard. The treadmill of production predicts that investors and producers will push for policies that expand markets and production despite the environmental consequences. E-waste is a classic example. In 1965, Gordon Moore, the founder of the Intel Corporation, predicted that the processing power of computers would continue doubling every eighteen months into the foreseeable future. "Moore's Law," as it became known, was an accurate prediction. Unfortunately, this increase in computing power was possible only with increased energy and natural resources use and more rapid cycles of obsolescence. One study revealed that between 1971 and 1995, as the computer chip evolved from earlier to more advanced models, the number of tran-

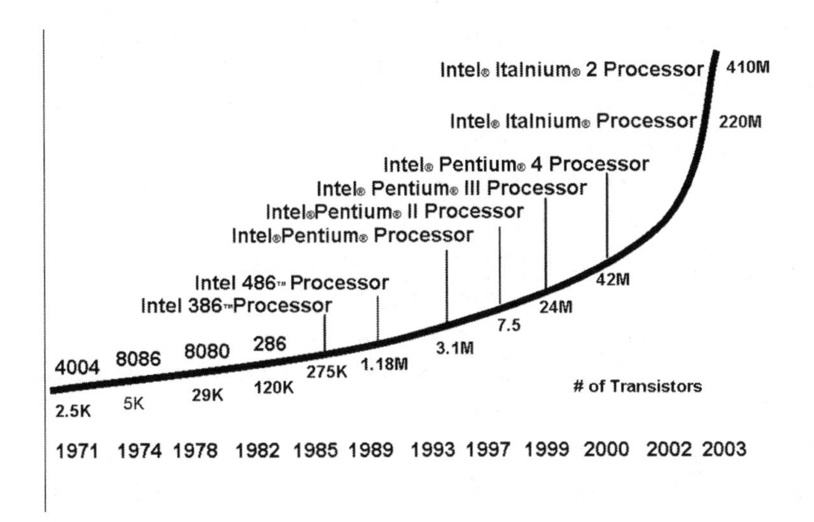

Figure 6.1
Moore's law. Source: Intel, *Moore's Law, The Future—Technology and Research*, http://www.intel.com/technology/silicon/mooreslaw/

sistors used in these machines jumped from 2,300 to 5.5 million (see figure 6.1).[13] This requires greater chemical inputs into production and ensures that each generation of computer products will become outdated sooner than the previous generation.

Computer sales are growing about 10 percent per year and more than 130 million computers are sold annually around the world, according to a recent report from the United Nations University:

By the end of 2002, one billion personal computers had been sold worldwide. Even though computers are becoming smaller and more powerful, their ecological impacts are increasing. The average 53 pound desktop computer with a monitor requires at least 10 times its weight in fossil fuels and chemicals to manufacture . . . which makes it much more materials intensive than an automobile or refrigerator, which only require one to two times their weight in fossil fuels to make. The natural resources extracted during the manufacture of computers are so intense that it is likely to have a significant impact on climate change and depletion of fossil fuel resources, according to the report. Because of this resource and energy intensity at the manufacturing stage of a computer's life, the energy savings potential of reselling or upgrading is some five to 20 times greater than recycling. The study notes that none of the EPR [extended producer responsibility] legislation in Europe encourages the extension of the lifetime of a computer, despite the huge environmental and economic potential.[14]

In another press report concerning the United Nations University study, the author noted, "The study criticizes governments for concentrating on recycling instead of introducing measures to reduce the numbers of new computers people buy, or encouraging them to buy secondhand machines. 'It's more effective to try and reduce and reuse things first and then worry about recycling' the study notes."[15] I would criticize environmental groups for taking the same approach, because recycling in a growth-based economy does little to challenge or slow the pattern of increased consumption and unfettered manufacturing. The dynamics of the computer industry would appear to follow a trajectory theorized by treadmill-of-production scholars rather than the model of ecological modernization.[16]

But ecological modernization is not just about the question of growth; it is also concerned with dematerialization—the use of fewer natural resources in favor of environmentally sound production. On this subject, a 2003 study suggested that information and communication technology offers few opportunities to dematerialize Europe's economy and could actually increase resource intensity. The findings challenge the industry's assertions that the sector is environmentally benign and an enabler of sustainable development. The report's author, Digital Europe, is an industry-backed research project examining the implications of e-business for sustainability. Even regarding information-based e-commerce, the report warns that significant economy-wide savings "'are not expected,' partly *because there are few products that can be reduced to an 'informational core'* and partly because consumer behavior is likely to change in ways that push resource use back up. The potential dematerialization 'appears to be small . . . [and the sector] might even be more resource intense than traditional retailing businesses.' A study of teleshopping, for example, showed no net transport savings. There are also indications, according to the report, that mobile teleworking and teleconferencing actually increase overall transport demand."[17]

Given the wealth of evidence that the production and disposal of electronics goods create toxic pollution within factories and communities, the Digital Europe study offers a look at the broader natural resource use associated with e-commerce. It seems clear that any hopes for ecological modernization might lie with another manufacturing sector, because the treadmill of production is hard at work in the world's most significant and dominant sector: the information technology (IT) industry.

Environmental Injustice and the Political Economy of E-Waste

The strong correlations among poverty, race, and pollution in the United States have been variously referred to as environmental racism, environmental inequality, and environmental injustice.[18] This is also a problem with a global reach. Like many other forms of pollution, e-waste also follows the path of least resistance and finds its way into poor nations, producing global environmental inequalities. Governments and industry leaders facilitate and allow these practices because they lead to profit making and the creation of low-wage work for many residents of economically desperate communities. The relative environmental privilege of northern nations is linked to and made possible by the environmental misery of southern nations in this regard. This reveals how global forms of environmental inequality take root and threaten communities but also how social movements engage the problem. Transnational social movement organization (TSMOs) frame the problem in terms of environmental injustice, explicitly acknowledging inequalities of race, class, and nation in the global distribution of industrial hazards.

An estimated 80 percent of computer waste collected for recycling in the United States is exported to Asia, where it is dumped and recycled under hazardous conditions.[19] Environmental activists have called this practice "toxic colonialism" and a form of "global environmental injustice."[20] Jim Puckett directs the Seattle-based Basel Action Network (BAN) and argues that the global trade in e-waste "leaves the poorer peoples of the world with an untenable choice between poverty and poison."[21] Puckett joined with other activists and organizations across thousands of miles to ensure that there is some accountability on the part of industries, governments, and consumers in the global North whose waste ends up in the South. He states, "This mentality perpetuated now by the United States is an affront to the principle of environmental justice, which ironically was pioneered in the United States and championed by the EPA domestically. The principle states that no people because of their race or economic status should bear a disproportionate burden of environmental risks. While the United States talks a good talk about the principle of environmental justice at home for their own population, they work actively on the global stage in direct opposition to it."[22] Figure 6.2 shows suspected

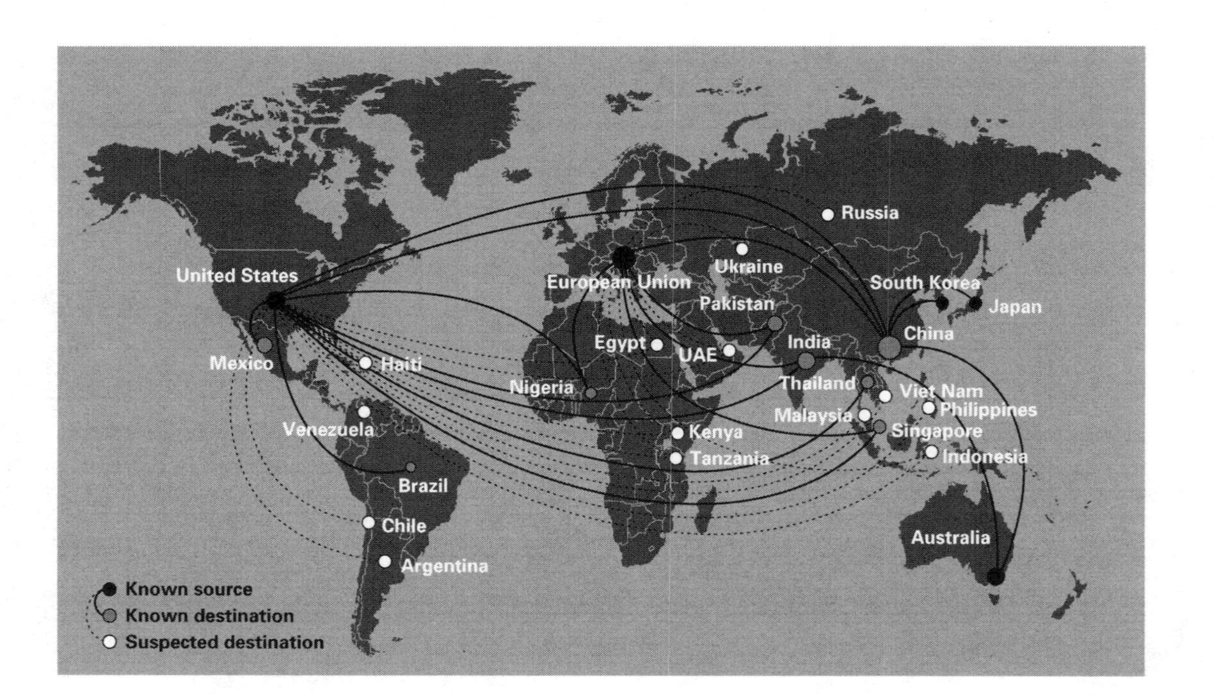

Figure 6.2
Known and suspected routes of e-waste dumping
Source: Dayaneni and Doucette (2005)

routes of global e-waste dumping, providing a visual sense of the traffic in high-tech trash.

U.S.-based activists became aware of the e-waste problem in the early 1990s, when journalist Bill Moyers produced a video and a book that revealed this phenomenon in China.[23] Shortly after, activists began to see the problem's roots in the United States. For example, Ted Smith of the Silicon Valley Toxics Coalition (SVTC) recalls:

> The e-waste problem emerged in the last few years of the USEPA's Common Sense Initiative [CSI]. There was one component of that that was looking at obsolete electronics. The industry wanted to deregulate e-waste. They wanted to get regulatory relief from having to treat old CRT [cathode ray tube] monitors as hazardous waste because of all the lead that was in it and . . . the EPA wanted to leave a large recycling loophole that says if it is recycled, then it doesn't have to be treated as hazardous waste. We were opposed to this, of course. It was through that process that I began paying attention to e-waste. It was growing. We were in a position to do something about this because we had been focusing on production and this was the end product, so it was logical. We all agreed to keep the materials domestic and not export any recyclables. But they [industry] actually shipped everything to China instead of doing that.[24]

Smith underscores the collusion between industry and government. Around the same time that the CSI group was meeting, the Public Interest Research Group in Delhi released a report on toxic wastes there, which concluded: "The information age has yielded the latest kind of waste. . . . The world's booming computer industry is creating computer waste ('Techno Junk') and this crisis is only beginning to be felt around the world."[25]

Responding to this problem, a transnational coalition of e-waste activists, the Computer TakeBack Campaign (CTBC), emerged in the late 1990s, determined to make the electronics industry more accountable to communities, workers, consumers, and the environment. The CTBC articulates an environmental injustice frame:[26]

> The recycling or direct dumping of the material results in a serious and immoral export of pollution to those countries [in the South]. Environmental protections in developing countries are usually poor, but regardless of the levels of protections, the export of pollution to countries due to their economic status is contrary to principles of environmental justice and moreover serves as a disincentive for manufacturers to prevent hazards and wastes upstream through product design. That is, rather than internalizing real environmental costs, manufacturers have been externalizing these costs to Asians and their environment.[27]

Activists from other SMOs that have worked on the e-waste problem concur. As a communiqué from the GrassRoots Recycling Network declared in response to the U.S. EPA's decision to allow e-waste exports, "Asian peoples are now asked to accept pollution that we have created simply because they are poorer."[28] Responding to similar reports, Von Hernandez a Philippines-based activist with Greenpeace International and the International Campaign for Responsible Technology, stated, "Asia is the dustbin of the world's hazardous waste."[29]

The transnational environmental justice networks that have evolved to track and combat the e-waste epidemic are clear in their framing of the problem as one rooted in inequalities by class, race, and nation and as perpetrated by both corporations and national governments. They articulate a model of global environmental inequality in a political economy that benefits consumers, private industry, and states in the North. Let us consider how this conflict plays out in the South.

E-Waste in the Global South

Although there are documented cases of e-waste dumping in Mexico, West and East Africa, and elsewhere, the majority of this waste ends up in Asia. And not only are states and corporations driving global environmental inequalities, environmentalists who strengthen regulations in the North unwittingly contribute to this problem as well. Ted Smith declared, "China's role as dumping ground for the world's unwanted gadgets is an outgrowth of efforts by wealthy countries to protect their own environments. . . . At the same time that we're preventing pollution in the United States, we're shifting the problem to somebody else. It's being exported and doing harm."[30]

Ravi Agarwal, an activist with Toxics Link India (an environmental justice group focused on e-waste in New Delhi), stated, "As developed countries become cleaner and it becomes very expensive to dispose of [e]waste because of rising environmental standards and labor standards, such waste finds itself in places like India and South Asia, and South East Asia."[31]

E-waste dumping is exacerbated because of the deep commitment of many Asian governments to developing and investing in electronics, as well as the success of many Asian corporations in that sector. In urban centers in India, Korea, Pakistan, China, Hong Kong, Vietnam, Malaysia, Taiwan, and the Philippines, there has been a marked growth in the development of a major infrastructure to support the electronics industry. Since so many of the electronics products sold on the global market are produced in Asia, e-waste dumping from consumer markets abroad back to Asia is viewed as logical, because the disassembled products can then be more efficiently incorporated back into production there. As Jim Puckett notes:

> A lot of people are saying [in response to our e-waste campaign], "Oh, but you know they are making a lot of this equipment now in China, so this is really a form of takeback [recycling], so we're just sending it back there." Doesn't that strike some kind of logical chord when you first hear it? But it is just amazing because basically you're saying, "We're going to take the dirtiest aspects of a product—which is the production side—and then, at the end of life—when it becomes a waste—send it to Asia. The middle part of the life cycle—when it is being used and it's benefiting people, it's being consumed and it's not very toxic at that stage, it's not putting out a lot of pollution—that's what we'll deal with in the West. But you in Asia can take it in the two dirtiest parts of the life cycle, and we'll justify it by calling it 'takeback.'"[32]

But if this is "takeback" or "recycling," then it is certainly not good for the environment.

E-Waste in the Philippines and China

The Philippines One of the earliest cases of resistance against e-waste dumping occurred when Greenpeace stopped a Russian-registered ship on its way to deliver forty-two tons of toxic computer scrap from Australia to the Philippines. The Greenpeace boat stopped the ship in Manila Bay before it could enter the port of Manila. A Scottish member of the Greenpeace team locked himself to the ship's anchor chain to halt the vessel. Simon Divecha, an Australian member of Greenpeace, declared, "Australia must stop sending hazardous waste to the Philippines. We want these containers of waste sent back to Australia." Greenpeace was originally seeking hazardous lead batteries, a common form of waste shipped to Asian lead smelters, but instead, they discovered containers of "used mixed electronic metal scrap" intended for "recycling."[33]

In the early 1990s, seventy-seven non-OECD nations and China pushed heavily for a ban on the shipping of waste for recycling purposes. As a result, in 1995 the Basel Ban was adopted in order to end the export of hazardous waste from rich OECD nations to poor non-OECD nations, even for recycling. The United States refused to participate in the ban. In fact, the United States has lobbied governments in Asia to establish bilateral trade agreements to continue dumping hazardous waste after the Basel Ban came into effect on January 1, 1998.[34] Hence, it should be no surprise that transnational export of e-waste remains a problem.

China Jim Puckett explained how BAN came to focus on the e-waste dumping and recycling phenomenon in China:

> The thing that really spurred me into it was I saw a . . . kind of online chat, when I did some word searching on the Web. It was a chat between recyclers, and one of them said, "Don't you think we should actually send an investigator over there, a private investigator to see what is going on, to see that this doesn't bite us in the butt later, because none of us really knows what's going on in China?" And they said, "Yeah, yeah, that would be a good idea," and then they never did anything. So that's when I thought maybe we should do that. We have some friends in the network in China, and I said, "Can you please spread the word that we're interested in this and if anybody has any information send me an e-mail." So all of a sudden I did get an e-mail that someone had read a news story in Chinese about an area that employed about 100,000 [people] along a river that was completely contaminated. So we found out about Guiyu, China that way. And we organized a trip out there.[35]

Ted Smith of SVTC was a key player in the BAN investigation effort in China:

> I was giving a presentation on e-waste at an international conference in New York. There were people from seventeen different countries there. They all signed on our "take it back" platform [a petition to urge companies and governments to push for electronics takeback policies]. That was what led to the decision to move forward with the *Exporting Harm* report. Several people who were at that meeting went back to their own countries and provided us info for the report. Without these opportunities, this wouldn't have happened.[36]

This activist work culminated in a 2002 report from BAN and the SVTC titled *Exporting Harm: The High-Tech Trashing of Asia,* which documented the growing international trade in toxic electronic waste from the United States to China, India, Pakistan, and other Asian nations.[37] Computer monitors, circuit boards, and other electronic equipment collected in the United States, sometimes under the guise of "recycling," are regularly sold for export to Asia, where the products are handled under hazardous conditions, creating tremendous environmental and human health risks. Workers, including children, use their bare hands, hammers, propane torches, and open acid baths to recover gold, copper, lead, and other valuable materials. What is unused is dumped in waterways, fields, and open trenches or burned in the open air.

Since 1995, the village of Guiyu in China's Guangdong province has been transformed from a poor, rural, rice-growing community to a booming e-waste processing center. Although rice is still grown in the fields, virtually all of the available building space is used for hundreds of e-waste recycling operations. Child labor is also widespread in the e-waste workshops in China. One immediate environmental impact has been the deterioration of the local drinking water supply. The local residents say that the water has become foul tasting, so they have clean drinking water trucked in from thirty kilometers away. Water and soil sample tests confirmed that lead and heavy metal contamination were much higher than levels allowed by the U.S. Environmental Protection Agency (EPA) and the World Health Organization.[38] The negative health effects associated with the work are in plain sight. A sixty-year-old resident of the region told a reporter, "For money, people have made a mess of this good farming village. . . . Every day villagers inhale this dirty air; their bodies have become weak. Many people have developed respiratory and skin problems. Some people wash vegetables and dishes with the polluted water, and they get stomach sickness."[39]

Subsequent studies and reports also concluded that e-waste laborers in China face serious health risks and the ecosystems affected by such operations are in significant danger.[40]

E-waste export is rooted in a simple, albeit brutal, calculus. As Ted Smith put it, "The reason is because it is ten times cheaper to send it to China than to recycle it here in the U.S."[41] E-waste shipments arrive at Hong Kong ports where officials are either too busy to inspect all containers or are sometimes bribed to allow the hazards to enter the country. The waste is then shipped to China for processing. BAN's Sarah Westervelt commented, "Brokers tell us they carefully tape $100 bills just inside the back of the shipping container so when the customs agents open up these containers, they've got their bribe and it can just pass on through." In some cases, the bribes are much greater than money: "The containers really are full of all sorts of things, including, we're told—though it's a little hard to believe—bribes as big as a Mercedes."[42]

According to one critic, e-waste dumping in Asia is a result of "America's NIMBY ["not in my back yard"] environmental policies."[43] But in Asia, another major factor is the quest for modernization, which pushes nation-states to invest in the electronics industry in the hopes of becoming an international economic juggernaut.[44] Thus, the northern desire for dumping grounds abroad goes hand-in-hand with Asian governments' desire to develop their economies. Commenting on global e-waste dumping from the United States to Asia, Chinese artist Danwen Xin produced a series of prints titled "disCONNEXION" that focus on this problem. Xin acknowledges the role of both the United States and China in creating this crisis:

> My goal is to sketch a visual representation of 21st century modernity. . . . I cannot forget that most of the e-trash I'm photographing is shipped from the United States and dumped in Guangdong Province, where people make a meager living recycling it. While we rely extensively on high-tech devices for our modern life, I was nevertheless shocked when I first confronted China's vast piles of dead and deconstructed machines, cords, wires, chips and parts—all with the traces of America on them. In my country, I have experienced the changes that have taken place under the influence of Western modernity. These changes, driven in large measure by the United States, have contributed to a strong and powerful push for development in China. At the same time, however, they have led to an environmental and social nightmare in remote corners of the country.[45]

The much-heralded Chinese economic superpower is both made possible and undermined by the social and ecological hell that residents and workers livie in, from coal mining towns to e-waste villages.[46]

E-Waste in South Asia and Africa

India Like China and other Asian nations, India has also embarked on a major modernization project through the embrace of information technology. But environmentalists and occupational health advocates are concerned the Indian government's priority of increasing computer density among the population will ultimately contribute to the waste problem. As in the case of China's imports of e-waste from the United States, brutal global economics plays a major role in India. Recycling a computer in the United States costs about twenty dollars, but the same product can be sold in New Delhi for a mere four dollars.[47] E-waste recycling also brings dangerous working conditions for those involved in disassembling these goods.

Toxics Link India activist Ravi Agarwal worries about the health impacts of exposure to plastics, for example: "When you actually physically break them [computers] down, which involves burning, putting them in acid baths, these very low end, basic, labor intensive breaking practices, then people breaking them get high degrees of exposure, [that are] totally unacceptable in most parts of the world."[48]

D. B. Boralkar, member-secretary of the Maharashtra Pollution Control Board and an expert on e-waste in India, notes that the burning of materials to extract metals for recycling is a "process [that] releases pollutants that cause diseases like silicosis, pulmonary edema, circulatory failure and suchlike."[49]

In *Scrapping the Hi-Tech Myth: Computer Waste in India,* Toxics Link India revealed that the "disposal and recycling of computer waste in the country has become a serious problem since the methods of disposal are very rudimentary and pose grave environmental and health hazards."[50] Hazardous waste import into India is actually prohibited by a 1997 Indian Supreme Court directive that reflects the Basel Ban on the prohibition of hazardous waste exports from OECD to non-OECD nations. Northern nations, however, continue to export e-waste to southern nations like India rather than manage it themselves. As a result, the trade in e-waste is camouflaged and is a thriving business in India, conducted under the pretext of obtaining reusable equipment or donations from northern nations.[51]

In 2004, the British Environment Agency (BEA) released a report indicting companies in the United Kingdom for sending tens of thousands of tons of e-waste illegally to India and other Asian nations. Kishore Wankhade of Toxics Link stated, "The trade is absolutely illegal and against the spirit of the Basel Convention."[52] His colleague Ravi Agarwal noted, "We have been repeatedly stating for the past two years that tons of e-waste are landing in various Indian ports every year for recycling. In the absence of access to customs data, this could never be verified. The BEA report, however, squarely indicts developed countries like USA and UK."[53] In 2004, K. S. Sudhakar, also of Toxics Link, found that a shipment of e-waste was mislabeled "metal scrap" when it arrived at the port in the city of Chennai, India. The mislabeling of toxics is one of the most common methods of getting such waste past the authorities and through loopholes in the Basel Convention. Toxics Link's research reveals that more than 70 percent of electronic waste collected in recycling facilities in Delhi was exported or dumped by northern nations such as the United States.[54]

Thus, from external and domestic sources, India has seen the health of its people, workers, and ecosystems deeply compromised by the cycle of electronics production and disposal.[55]

Africa All over Africa, as in many inner-city school systems in the United States and elsewhere, one way that e-waste dumping occurs is through the stealthy practice of "donations" or "charity."[56] Firms receive tax write-offs and do not have to pay for disposal of computers they donate to nations or various charitable causes. Von Hernandez of Greenpeace calls this a scam because firms not only save money, but "they get rid of the computers right before the point at which they would normally be sent to the scrap yard. So instead of having to process them and have them recycled, they send them to another nation and sooner or later they'll get dumped there."[57] This practice is what activists call delayed dumping (see chapter 4).

Well-meaning nonprofit organizations like Computer Aid International (CAI) ship older computers to schools in South Africa, Uganda, Kenya, and Nigeria with the intention of bridging the digital divide and providing access to this technology for some of the world's poorest children. Tony Roberts, head of CAI, estimates that 99 percent of schoolchildren in global South countries leave school without touching or seeing a computer in the classroom, and he would like to change that.[58] While groups like CAI are careful about testing computers for safety,[59] many computers shipped from places such as the United Kingdom to points in Africa for school use have been declared useless on arrival because of incompatible software configurations and the need for maintenance on already failing machines.[60] Wilfred Baanabakintu, the public relations officer for the National Environment Authority in Uganda, stated, "In the guise of giving us aid . . . there's a tendency of dumping here computers and this is a probably because of our poverty . . . we are vulnerable to waste from these hi-tech countries."[61]

This is high-tech environmental inequality, as the majority of these toxics flow from North to South in the name of development. Unfortunately, even with the EU's Directive on Waste from Electrical and Electronic Equipment (requiring manufacturers to recycle electronics products at the end of life), if producers are not actually taking responsibility for recycling products domestically, many experts expect to see a continued rise in the export of hazardous e-waste to Africa, Pakistan, India, China, and elsewhere in the South. In 2003, record volumes of e-waste—23,000 tons of it—left Britain for such destinations. And the government figures indicate that more than ever before, these materials are being shipped abroad. This is particularly acute since in Britain, electronic goods are required to be recycled and barred from incinerators and landfills.[62]

In October 2005, BAN released a report documenting a major increase in e-waste exports from Europe to West Africa. For years at Basel Convention meetings, representatives from African nations had protested that e-waste dumping was on the rise in the region, but that UN body paid little attention to those claims. Finally, BAN and its partner organizations carried out an investigation and found that West African nations like Nigeria are becoming hosts for extraordinary volumes of e-waste, creating human health and ecological risks throughout the region. The major difference between e-waste dumping in West Africa and Asia is that the latter regions have at least some of the infrastructure to recycle and remanufacture these products. BAN confirmed what many observers expected: West African economies have little capacity to properly recycle electronics products. Moreover, confirming activists' suspicions that these exports are more about dumping waste than feeding global South recycling operations, the Nigerian Computer Dealers' Business Association informed BAN investigators that as much as 75 percent of the imported electronics material received in that nation is beyond use and unsalvageable.[63] As with previous eras of foreign municipal and chemical hazardous wastes, West Africa continues to serve as a dumping ground for the global North.

E-Waste in Latin America

Although data are sparse, there is evidence that e-waste is making its way to points in Latin America. The *maquiladora* region along the Mexico-U.S. border is particularly problematic. Diane Takvorian, executive director of the Environmental Health Coalition, a binational environmental justice group, observed: "We've seen a growing number of dirty recycling and metal recovery operations along the Mexican border. We already see elevated levels of lead, mercury and other heavy metals. Shipping millions of tons of unregulated e-waste across the border poses a serious threat to Tijuana and San Diego environmental quality and public health."[64] And because there are many electronics firms in the maquiladora region, they create their own significant volume of e-waste, although under absentee ownership, so the financial benefits accrue to foreign firms while the environmental costs are borne by the local populations. In fact, Tijuana, Mexico, hosts the most television-producing firms anywhere, earning it the title of "TV capital of the world."[65]

There are few reports of e-waste dumping in other parts of Latin America, but that could change as rapidly as we have witnessed other waste flows shift from one part of the global South to another, depending on resistance from TSMOs and governments and the contours of waste markets. As with many other parts of the world, Latin America's middle class is increasingly using cell phones and computers, generating e-waste from within the domestic sphere. Recent reports of major spikes in e-waste production from within Chile have that nation's environmentalists and government agencies worried about the long-term impacts of such excesses.[66]

There are three primary reasons why e-waste is increasingly flooding southern nations:

- Labor costs are very low (in China e-waste workers earn only $1.50 per day).
- Environmental and occupational regulations are lax or not well enforced.
- It is legal in the United States (despite international law to the contrary) to allow export of hazardous e-wastes with no controls whatsoever.[67]

Politically, ideologically, and culturally, e-waste environmental injustice is sanctioned because poor people in general and people of the South in particular are valued less than peoples in the North, so the level of concern and sensitivity that might otherwise prevent this dumping is absent from international politics. But there is reason for cautious optimism, as movements and states are acting.

As local governments around the United States have seen this new waste disposal problem emerging, they have begun to sound the alarm. Governments have taken on the burden until now, but states and municipalities are arguing that corporations should bear more of the costs. Michael Alexander, a senior research associate with the National Recycling Coalition, points out, "The question being raised everywhere is: Should local government be straddled with this cost? And shouldn't manufacturers be involved? The question of manufacturer responsibility is now coming to the forefront."[68]

The Global Movement for Extended Producer Responsibility

Many TSMOs are based in urban centers of the global North. This is strategic in large part because so much of the global political and economic power and decision-making authority rests in those spaces and in institutions located there. Thus, one would likely have a greater chance of changing a transnational corporation's practices in many global South nations by targeting that firm's headquarters in the North rather than focusing on its operations in the South. This is precisely what TSMOs have done by targeting corporate practices directly or through government legislation. This is a political economic opportunity structure approach that activists have adopted to strategically apply pressure wherever the decision-making power rests, whether within state apparatuses or corporate boardrooms. The grassroots pressure that TSMOs mobilize against electronics firms in particular comes not only from their own activist membership and staff but also from the consumers of these technologies. Given that such a high percentage of citizens in the North are consumers of computer and electronics products, this provides a considerable pool of potential activists who might be mobilized to pressure or even boycott any number of companies.

Four U.S.-based SMOs have been critical to the success of various campaigns to reform electronics manufacturer practices: the Basel Action Network (BAN), the GrassRoots Recycling Network (GRRN), the Silicon Valley Toxics Coalition (SVTC), and the Texas Campaign for the Environment (TCE). Together, these SMOs created the Computer TakeBack Campaign (CTBC), a U.S.-based effort to make computer producers responsible for their products at the end of life. Two of these organizations (SVTC and BAN) are actually TSMOs and have contributed to successful efforts to pass such legislation in the European Union (EU). They also work closely on e-waste recycling campaigns with SMOs in other nations, including the Clean Production Network (Canada), Greenpeace International (the Netherlands), Greenpeace China, Toxics Link India, Shristi (India), SCOPE (Pakistan), and the International Campaign for Responsible Technology (ICRT).

The mission statements of each of the U.S.-based leading SMOs on e-waste issues reveal their understanding that environmental justice must be approached through the intersection of markets and politics. The following statements underscore this framing. We begin with the two SMOs:

Texas Campaign for the Environment is dedicated to informing and mobilizing Texans to protect the quality of their lives, their health and the environment. We believe that people have a right to know and a right to act on issues that fundamentally affect our lives and future. TCE cannot compete with *corporate polluters when it comes to writing checks for the election campaigns of politicians who make the laws*. However, we win when we organize at the grassroots level and gain strength in numbers from people like you who get involved and support our work.[69]

[The GrassRoots Recycling Network's] mission is to eliminate the waste of natural and human resources—Zero Waste. We utilize classic activist strategies to achieve *corporate* accountability for and public policies to eliminate waste, and to build sustainable communities. . . . *We prioritize corporate accountability because global corporations are the primary engines of environmental and social destruction.* The key tool to achieve corporate accountability for waste and ultimately Zero Waste, is extended producer responsibility (EPR)—the principle that manufacturers and brand owners must take responsibility for the life cycle impacts of their products, including take back and end of life management.[70]

The two TSMOs adopt a similar framing of the issues:

Silicon Valley Toxics Coalition (SVTC) is a diverse grassroots coalition that engages in research, advocacy, and organizing around the environmental and human health problems *caused by the rapid growth of the high-tech electronics industry.*[71]

[The Basel Action Network (BAN) is] an international network of activists seeking to put an end to economically motivated toxic waste export and dumping—particularly hazardous waste exports from rich industrialized countries to poorer, less-industrialized countries. Unfortunately, very *powerful governments and business organizations* are still trying to overturn, circumvent or undermine the full implementation of the Basel Ban and in general seek to achieve a "free trade" in toxic wastes.[72]

A much broader umbrella network to which all these groups belong is the ICRT, an international solidarity network that promotes corporate and government accountability in the global electronics industry. Ted Smith, one of the ICRT founders, explains how the network began:

It was the late 1980s and early 1990s when we first got interested in the international angle of e-industry. At that time industry was moving out of Silicon Valley, particularly to the Southwest U.S., so we started working with SNEEJ [the Southwest Network for Environmental and Economic Justice] in particular and other groups in the United States. But then it was pretty clear it was beyond that—internationally. We began to hear from people in other countries, and we started to reach out to them. They started to deal with these aspects, and we started making linkages and we started meeting at conferences. It was basically the same sets of issues—groundwater contamination and worker health, deregulation and corporate welfare—giving away of huge subsidies. In 1990 SVTC formed the Campaign for Responsible Technology, which later became the International CRT.[73]

Drawing on the ICRT's resources, these movement groups and networks collaborated to create the CTBC:

The goal of the Computer TakeBack Campaign is to protect the health and well being of electronics users, workers, and the communities where electronics are produced and discarded *by requiring consumer electronics manufacturers and brand owners to take full responsibility for the life cycle of their producers, through effective public policy requirements or enforceable agreements.*[74]

Separately and in combination, these social movement organizations, networks, and their campaigns focus on decision makers in governments and business organizations, revealing a broader political economic opportunity structure model of movement targets. The centrality of the links between corporations and the state is critical here. These links are most visible and effective in urban centers like Austin, Boston, San Francisco, Seattle, Silicon Valley, and, increasingly, points in Asia, Europe, and Latin America.

One of the southern partners in the global campaign to combat e-waste is Toxics Link India. Toxics Link works with ICRT, SVTC, BAN, and other northern groups to highlight the growing e-waste crisis, and its mission underscores the political economic opportunity structure approach:

Toxics Link's goal is to develop an information exchange mechanism that will strengthen campaigns against toxics pollution, help *push industries towards cleaner production* and link groups working on toxics issues. We are a group of people working together for environmental justice and freedom from toxics. We have taken it upon ourselves to collect and share both information about the sources and dangers of poisons in our environment and bodies, and information about clean and sustainable alternatives for India and the rest of the world.[75]

In addition to emphasizing the corporate role in the global e-waste crisis, Toxics Link is explicit about the need to "link groups working on toxics issues" around the globe, because its representatives know that transnational environmental justice problems require coordinated transnational action.

Another Asian SMO working with these organizations is the Taiwan Environmental Action Network (TEAN). TEAN activist Wenling Tu stated:

TEAN has been working with local grassroots organizations, documenting the most pressing environmental and social impacts associated with high-tech development in Taiwan since 1999. As Taiwan has established many IT clusters nationwide, it becomes pressing to solve the issue of high-tech development at root. In the past 5 years, TEAN, along with local communities and international NGOs, advocates environmentally sound practices and social responsibility of IT firms and demand the government to address the high-tech environmental impacts.[76]

These U.S.-based movement groups and many of their international partners joined forces to produce a report and video documentary, *Exporting Harm: The High-Tech Trashing of Asia.*[77] After months of strategizing with SMOs to gain access to sensitive sites and interview workers in China, Pakistan, and India, they completed and released the report, which sent shock waves through the electronics industry and was picked up by nearly every major media outlet in North America, Europe, and Asia. The struggle to prevent the export of waste to the global South was critical, so activists knew that this effort had to begin in the North.

Hence, we examine the most important computer waste campaign to date: the action directed against the U.S.-based Dell Corporation.

The Dell Campaign

The CTBC emphasizes extended producer responsibility (EPR), the emerging global framework that holds producers and brand owners financially responsible for the life cycle impacts of their products, with particular emphasis on product take-back and end-of-life management. By shifting the costs of managing discarded products away from taxpayers and local governments onto producers and brand owners, EPR creates a powerful market incentive for manufacturers to reduce those costs by redesigning products to be less toxic, more reusable, and more easily recycled. The CTBC goal statement captures this idea well: "Take it back, Make it clean, and Recycle responsibly." Since its inception, the CTBC has pursued a dual strategy, which reflects a political economic opportunity structure approach: build sustained consumer and market pressure on computer corporations, and build informed public support for regulatory reforms embracing producer responsibility. The CTBC's founding organizations quickly recognized the power inherent in a markets campaign (sometimes called corporate campaigns), paired with a policy campaign focused on state legislative change.

Prior to the Dell campaign, activists achieved an important victory, which laid the groundwork for the action against the firm. In 2001, the U.S. EPA launched the National Electronic Products Stewardship Initiative (NEPSI), a multistakeholder dialogue on resolving the e-waste problem in the United States. By urging companies to adopt reuse and recycling initiatives, NEPSI hoped to avoid angering industry with new regulations while giving environmentalists a voice at the table. A split soon emerged between companies that wished to develop a back-end financing scheme (called an advanced recycling fee, whereby consumers pay a fee to firms that recycle their old computers) and companies preferring front-end programs (where the firms pay the costs). Representatives from major television manufacturers and IBM supported the back-end proposal, while Hewlett-Packard (H-P) and environmentalists supported the front-end approach. IBM and television industry representatives opposed the front-end idea because they were the largest producers of historic electronic waste and would therefore have to bear the greatest burden if consumers began returning old products.[78] This split in the NEPSI group eventually contributed to that initiative's collapse, in large part because environmentalists exploited this division within the industry. The political economic opportunity structure framework views instabilities among corporate elites as just as important as disunity among political elites (the latter being a primary emphasis of the traditional political opportunity structure model; see chapter 2). This division allowed environmentalists to push some of the most influential firms toward more progressive practices with respect to product recycling.

Dell's business model is fundamentally about a relationship with its customer base that uniquely positions the company within the computer industry to develop and successfully implement a comprehensive, national take-back system. Dell is the only computer company in the United States (if not the world) that knows all of its customers by name, mailing address, e-mail address, telephone number, date of purchase, and product specification—exactly the kind of information a company would need to design a system to recover its obsolete products.

Furthermore, Dell is not so much a manufacturing company as it is a marketing company. That is, it assembles made-to-order computers from parts supplied to it and attaches its logo. CTBC activists believed that Dell, as a marketing company, was particularly susceptible to a strategy that attacked its brand name. The company bears the name of its founder, Michael Dell, who is the CEO, a major stockholder, and the corporation's most visible personality. This arrangement provided CTBC the opportunity to personalize the issue and associate it with Michael Dell, who takes credit for the company's direction and success. The CTBC issued a statement declaring, "The Dell position on e-waste is a stain on the soul of Dell—the company and its founder, . . . Michael Dell and his wife, Susan, make generous donations to children's health and environmental charities in the United States but ignore the health and environmental impacts of e-waste on children and adults."[79] In a humorous effort to reach a younger audience of students and other computer consumers, the CTBC referred to Mr. Dell as the "Toxic Dude."[80] The CTBC was also able to secure face-to-face meetings with Michael Dell on at least four occasions in 2002 and 2003 at shareholders' meetings and at the company's headquarters in Austin, Texas.

The CTBC issued "report cards" that tracked Dell's progress on recycling and was effective at capturing press headlines and getting consumers involved. The report cards pit Dell against its competitors like H-P, which was doing a better job of recycling its products, although still doing very little. In its 2003 annual report card, the CTBC assigned failing grades to Micron Technology, Gateway, H-P, and others, and placed emphasis on the shortcomings of the Dell Corporation.

Dell unwittingly provided the CTBC with other targeting opportunities, including the company's decision to partner with UNICOR, the federal prison industries, as its primary recycling partner. UNICOR is a publicly subsidized prison industrial operator that runs a chain of plants in the federal penitentiary system. Activists visited a prison in California where Dell contracted for this kind of work. Ted Smith and Sheila Davis toured the prison. Davis, director of SVTC's Clean Computer Campaign, stated, "We were appalled to witness the working conditions inside the federal prison at Atwater, California, where inmates were using hammers to smash computer monitors."[81] One of the prisoners commented: "I work just outside [the cathode ray tube glass breaking area], and am offered no ventilator, and they won't give us blood tests. It's a Mickey Mouse operation, and inmates are knowingly being subjected to chemical cocktails—and that is the bottom line. We are guinea pigs and slaves, and treated precisely that way."[82] Another inmate stated: "Funny, isn't it, how this stuff is unsafe for public dumps, but not for us lowly prison inmates?!? Quite the double standard, wouldn't you agree?"[83] Another e-waste worker at the prison in Atwater reported, "Even when I wear the paper mask, I blow out black mucus from my nose every day. The black particles in my nose and throat look as if I am a heavy smoker. Cuts and abrasions happen all the time. Of these the open wounds are exposed to the dirt and dust and many do not heal as quickly as normal wounds."[84]

Prison inmates reported that those who sought to improve conditions in the e-waste recycling facility faced discipline and the threat of job loss. Inmates earned from $0.20 to $1.26 per hour at the Atwater prison.[85] These workers toil outside the protection of environmental and labor regulations that private sector employers must obey, including federal Occupational Safety and Health Administration standards.[86] The laborers are not classified as employees and are not protected against retaliatory acts by their employer under the Fair Labor Standards Act. Inmates are not allowed to unionize or serve on the prison health and safety committees.

For Dell, selecting UNICOR was a matter of driving down costs. But the CTBC's concern was that reliance on prison labor placed these workers' health at risk, and it undercut development of the market infrastructure necessary to operate a robust, national e-waste collection and recycling system. Activists, using an environmental justice frame, charged that Dell's recycling program was "a high tech chain gang" and not much better run than the e-waste sweatshops in Guiyu, China.[87]

To dramatically drive home the point, a group of CTBC activists dressed in prison uniforms circled a mound of dusty computers outside the national Consumer Electronics Show in Austin, Texas, in January 2003. Linked together by chains, these mock prisoners wore "Dell Recycling Team" signs and chanted, "I lost my job. I robbed a store. Went to jail. I got my job back." The campaign continued.

David Wood, former codirector of the GrassRoots Recycling Network, was critical of the prison recycling program:

> These recycling operations suggest two divergent paths for the future of e-waste recycling in America. One path leads toward efficient, transparent, modern facilities staffed by free labor, possessed of their rights as contemporary employees, able to protect themselves and nearby communities from harm. The other path descends into a closed, Dickensian world of prisoners condemned to dangerous work for little pay under backward conditions. Depending on the path we choose, e-waste recycling can contribute to community economic development and environmental protection, or can become the equivalent of breaking rocks on a high-tech chain gang.[88]

In June 2003, the CTBC released "A Tale of Two Systems," a report that contrasted Dell's prison-based recycling operations against H-P's free market partnership with the firm MicroMetallics. Barely a week after the CTBC released this report on its concerns about the use of prison labor, the Dell Corporation announced that it would no longer rely on prisons to supply recycling workers for its program.[89] The fact that the Dell Computer Corporation cancelled the contract one week after the CTBC made its objection to these practices public is a testament to the real power this activist network exercises against global corporations.

Activists remained relentless in their public criticism of the Dell Corporation and its founder. One way of approaching this problem was to question the existence of Dell's dual recycling system—one for the United States and one for Europe. The CTBC received data and details on this "double standard" from its partner SMOs across the Atlantic and made it known publicly. As Robin Schneider, director of the Texas Campaign for the Environment, stated:

> We want Dell Computer to take the same degree of responsibility for used and obsolete personal computers here in the United States as the company does in European countries. In Europe, Dell takes back old equipment free of charge from all consumers. European producer responsibility laws require this of the company, ensuring that the products are kept out of landfills and incinerators and their valuable materials are reused or recycled. Our simple question to Dell Computer is "Why do American consumers and the American environment deserve second-class treatment?"[90]

In July 2003, activists leafleted the Dell shareholders meeting in Austin, Texas, and delivered truckloads of Dell computers collected from several western cities as part of an action creatively labeled the "hard drive across the West." CTBC activists drove trucks and held events in five cities in the western United States, picking up obsolete computers along the way and, using innovative guerrilla street theater to capture media headlines, forced Dell to take notice. The Dell corporate staff accepted the waste computers and even helped activists load them onto a truck to be hauled to Image Microsystems, an Austin-based recycler that replaced UNICOR, which had previously managed Dell's waste. Schneider observed, "We're not exactly holding hands and singing 'Kumbaya,' but this is more common ground on recycling than we've had for a while." Image Microsystems also agreed to honor CTBC's Recycler's Pledge of Stewardship. Included in the pledge is the requirement that toxic waste not be sent overseas to global South nations or dumped in domestic landfills. Ted Smith stated, "We consider this a big win. But we're also very mindful that Dell only recycled 2% of the 16 million computers they sold last year. Those uncollected computers will still end up in landfills or shipped overseas."[91] In 2004, the CTBC achieved a key victory by persuading Dell to join H-P in endorsing a statement of principles in support of producer responsibility. The CTBC urged Dell to go further because it was the largest computer company in the world and because it wanted Dell to bring its policies into alignment with H-P, its strongest competitor. This action allowed the CTBC to capitalize on competition within the industry, creating a powerful point of leverage within the political economic opportunity structure.[92] Consider what the CTBC had accomplished thus far: in less time than it takes to introduce proposed legislation to a city council, this group of SMOs forced the largest computer company in the world to cease the use of prison labor, forced the company to choose another contractor altogether, persuaded that contractor to adopt the CTBC's pledge of stewardship and principles of environmental justice, and persuaded Dell and its strongest competitor firm, H-P, to endorse a pledge of producer responsibility. These were remarkable achievements.

Consumers were critical to the Dell campaign. The CTBC persuaded thousands of Dell customers and students to send e-mails, letters, and faxes to Michael Dell, urging him to institute a recycling program that would take back computers and recycle them without prison labor. Student groups were vital in this effort, having raised funds to place an open letter in a full-page advertisement in the *Austin Chronicle,* a major newspaper in Dell's hometown. This letter represented 153 student organizations from all fifty U.S. states. Students also successfully negotiated a meeting with Michael Dell (which was webcast live), where he responded to a list of concerns and demands.

In 2004, Dell and H-P, the nation's largest personal computer makers, reported that they were moving to support more recycling and taking more of the financial burden for recycling used computers off consumers and local governments. These pledges were timed to the release of another annual CTBC report card on corporate environmental behavior. H-P received the highest rating, and Dell moved up to second place on the 2004 report card. Ted Smith stated, "We believe the companies have to set up these systems, not governments."[93] Even so, the CTBC continues to push state governments to address e-waste recycling needs. During 2003 alone, more than half the states in the United States introduced electronic waste legislation.[94] This is testimony to the dual-pronged political economic opportunity structure approach that activists use to ensure that states and corporations behave in ways that approach environmental justice.

Much of the waste from companies like Dell, H-P, and others is exported to the global South, creating further social and ecological disruptions. E-waste recyclers face significant health hazards while disassembling the waste from global North and domestic middle-class consumers. They too are part of the chain of production, consumption, and disposal, and CTBC activists were clear that they deserve our attention, access to basic rights, environmental protections, and a living wage. After all, without ties to workers and environmentalists from social movement organizations in these southern nations, the information about e-waste dumping might never have surfaced.

Continuing E-Waste Struggles in the South

The primary responsibility for global environmental inequalities lies with states, corporations, and consumers in the North. But activists in the South are working hard to confront the situation on the ground in their own countries as well. Recycling may appear to be ecologically responsible, but e-waste recycling is anything but that. And communities in the global South are resisting it as fast as communities in the North are embracing it.

In March 2004, Citiraya Industries Ltd. proposed to build an electronics recycling plant in Melaka, Malaysia. That same month, it announced that it was investing $65 million to do the same in China. The Singapore-based company also has a plant in its home nation, where it processes fifteen thousand tons of electronics annually.[95] But as rich nations cast off more of these products every day, the volume is growing at a breakneck pace, so Citiraya is meeting the demand. However, the plant in Malaysia was never built. As Mageswari Sangaralingam, a research officer for the Consumers' Association of Penang, Malaysia (CAP), told a group of fellow activists on a global activist listserv, "Citiraya's proposal to build an e-waste recycling plant in Melaka, Malaysia was rejected by the State government due to public protest (although the Environmental Impact Assessment was approved by the Department of Environment). So it seems like the company is looking for new frontiers where there may not be objections from locals"[96]

Activists deemed the proposed plant a hazard to human health because it was to be located in an area of high population density and near businesses producing food and pharmaceutical products. Citiraya initially proposed the construction of a waste incinerator in 2003, "but after strong opposition," they changed plans to reflect a softer approach of recycling.[97] State Health, Manpower and Consumers Affairs committee chairman Datuk Seah Kwi Tong said, "The government recognizes the views of the people. It also recognizes that human error could happen and it might have fatal consequences to a densely populated area."[98] Acknowledging the cooperation and assistance of environmental justice activists from various nations, activist Sangaralingam noted, "Thank you all for your help in providing comments and locating resource persons for expert input on the review of the EIA [environmental impact assessment]. Your assistance in this matter is deeply appreciated."[99]

Later, in an interview with Sangaralingam, she gave more details of her organization's involvement in this conflict:

> The Consumer's Association of Penang (CAP) had raised our objections in October 2002 and our comments were incorporated in the EIA of the proposed project. The factory, which has e-waste recycling plants in Taiwan and Singapore, planned to crush and grind the e-waste with a hammer mill and also extract the precious metals through a solvents extraction and dissolution process. We had the

assistance of several foreign experts who gave input and valuable comments on the technology. We shared this information with the community objecting to the proposed plant and on December 31, 2003, we were informed that the State government has decided not to approve the e-waste plant in that location due to public protest and objections from other parties although the EIA of the proposed plant was approved.[100]

Rather than adopting a NIMBY approach to toxic facility siting, CAP articulated a vision of sustainability and environmental justice that locates the source of problems and solutions with the state and industry. Sangaralingam stated:

We are of the opinion that electronic waste recycling is a commendable effort but it has to be done safely as the components are mostly hazardous in nature. The authorities should look into the suitability of the site and effectiveness of the pollution control technologies. A clean production approach, which substitutes safe materials and processes to stop the generation of hazardous e-waste is needed. We are urging the Malaysian government to expedite the development of e-waste legislation in the country. Efforts to set-up a domestic recycling facility must be coupled with extended producer responsibility (EPR) legislation.[101]

In other parts of Asia, e-waste politics rages on. For example, as Taiwan's electronics industry continues to grow, activists with TEAN and Greenpeace International are reminding companies there of the realities of e-waste and urging large Taiwanese firms like Acer to keep their promises to reduce toxic inputs. In June 2006, Greenpeace and TEAN held a protest at the Computex trade show in Taipei, Taiwan. Nine campaigners wearing protective suits and masks stood in front of an entrance to the Taipei World Trade Center and displayed posters with pictures of children handling e-waste. Beijing-based Greenpeace activist Jamie Choi stated, "We wanted to use Computex Taipei as an opportunity to let the Taiwanese industry know that they are using toxic substances inside their products and we want all the industry, not only the Taiwanese industry, to stop using toxins inside their products."[102] Several companies have agreed to time lines to phase out toxic substances in their products, including H-P, Nokia, Samsung, Sony, and Sony Ericsson. Greenpeace applauds these promises and is putting pressure on other large firms such as Apple, Dell, Fujitsu Siemens, Lenovo, Motorola, Panasonic, and Toshiba to follow suit.[103]

Greenpeace focused particular attention on H-P since it is one of the world's largest computer firms and has historically included a range of toxics in its products, which means a lot of H-P's e-waste ends up in Asia.[104] In May 2005, the group launched a public campaign directed at H-P, when activists showed up at the Eighth China Beijing International High-Tech Expo and unveiled a 2.7 meter high sculpture of a wave at the event's opening. The sculpture was created from e-waste gathered in Guiyu scrap yards. At that same moment in Geneva, Greenpeace activists delivered a truckload of e-waste to H-P's headquarters, declaring H-P a "toxic tech giant," demanding that the company clean up its act, and urging other firms to do so as well.[105] Later that year, in December, Greenpeace activists protested at H-P's offices in Palo Alto, California, further elevating the campaign's visibility and raising the company's public relations costs of ignoring these demands.

By early 2006, Greenpeace had achieved victory: H-P announced that it would phase out BFRs and PVC from its products.[106] Greenpeace China activist Kevin May went to H-P's shareholder meeting in March 2006 to thank the company for its commitment and to push it to phase out even more toxic substances in its products.[107]

Greenpeace India conducted a similar protest action in Bangalore, India, in September 2006, when it dumped five hundred kilograms of Wipro-branded e-waste at that firm's corporate headquarters. Wipro is an Indian-based IT firm that has become a major global corporation, with offices in numerous nations. As India's growing middle class increasingly uses, consumes, and disposes of more computers, this creates a serious domestic e-waste problem. The protest was based on Greenpeace's claim, supported by the Karnataka Pollution Control Board, that Wipro was sending significant volumes of e-waste to unauthorized recycling sites.[108]

The phenomenal economic growth in India and China, coupled with computer production there and in other Asian nations like Japan and South Korea, underscore that North-to-South environmental inequalities occur regionally and domestically—not just between the United States and Europe versus the rest of the world—and are as much about economic hierarchies as they are about ethnic and national inequalities. In fact, environmental inequalities within Asia result from this process, as Greenpeace China's and BAN's investigation in Taizhou, China, in 2004 discovered large volumes of e-waste dumped there as a result of exports from South

Korea and Japan.[109] Hence, the term *global South* can apply to vulnerable communities in the United States and Europe just as it applies to similar communities in Asian, African, and South American regions and nations.

Activists in the South—in Asian nations in particular—are working to prevent indiscriminant e-waste dumping in their nations and communities. Given the combination of overproduction and consumption in the North, with the general embrace of electronics technology by many Asian states, this has been difficult.

State Responses

From local municipalities to national and international bodies, governments from California to Asia have implemented EPR legislation to address the e-waste crisis and activist demands. However, the major obstacle remains the U.S. White House and U.S.-based trade organizations representing the electronics industry.

The major regulatory control on hazardous waste flows from global North to South is the Basel Convention and the Basel Ban, the international agreement prohibiting the export of hazardous waste from OECD to non-OECD nations, even through recycling. In principle, this ban should work, although in practice, industries continue to send waste to the South illegally. One major problem is that the United States, the world's most influential economic power, has yet to ratify the agreement (only two other governments have refused to ratify the Basel Ban—Afghanistan and Haiti). Environmental activists say that aside from its position on Basel, the U.S. government's behavior is even more problematic: "The U.S. government policies appear to be *designed* to promote sweeping the e-waste problem out the Asian back door. Not only has the U.S. refused to ratify the Basel Convention and Ban, but also in fact, the United States government has intentionally exempted e-wastes, within the Resource Conservation and Recovery Act, from the minimal laws that do exist (requiring prior notification of hazardous waste shipments) to protect importing countries."[110]

The U.S. White House and Congress refuse to do anything constructive about the e-waste problem for now. But other governments have. Immediately after the "Exporting Harm" report was released, the government of China responded with stepped-up inspections and arrests at its major ports and a declaration that it will not accept e-waste smuggling. In a series of media releases in the wake of the "Exporting Harm" report, the Chinese government called on exporting nations to take responsibility for their own electronic waste.[111] Realistically, this may be largely symbolic politics given the countervailing drive by Chinese industry and the state to modernize the economy, which has achieved an annual growth rate of 8 percent in recent years.[112]

Individual states within the United States have responded as well. In September 2003, the governor of California signed landmark legislation establishing a funding system for the collection and recycling of certain electronic wastes. Maine and at least twenty-five other states are following suit with proposed laws to have producers take back e-waste. In 2006, Washington State's legislature passed the most comprehensive e-waste bill to date, which requires manufacturers to finance the collection and recycling of electronics products at no additional cost to consumers. Ted Smith, who was involved in the Washington State effort, observed: "There is legislation in several states that is pushing forward extended producer responsibility at the state level, efforts to bring about this kind of a solution. The industry is trying to get a national solution to pre-empt the state efforts. I met with a U.S. senator's aides who wants to hold hearings on the exports issue."[113]

At the international level, the most progressive state action yet is in the EU, which passed two major policy initiatives, known as the Directive on Waste from Electrical and Electronic goods (WEEE) and the Restriction of Hazardous Substances (RoHS). These policies require electronics producers to take back products at the end of their life and reduce the use of toxics in production. This should allow products to be recycled and reused instead of being dumped into landfills or exported. Individual European nations are also moving forward with progressive policymaking and action on the problem of e-waste export. For example, in 2004, the British Environment Agency (BEA) concluded that in response to reports of large volumes of e-waste being shipped from the United Kingdom to non-OECD nations in Asia and Africa, it would step up inspections at ports with a newly formed dedicated special enforcement team and would coordinate with twelve other EU nations to do the same across twenty-five international ports.[114] But, like China's inspections, these are also short-term actions.

Not only are the U.S. White House and Congress resisting progress on e-waste management, they have been hard at work hindering others from doing anything positive about the situation as well. For example, on behalf of the American Electronics Association, the U.S. Trade Representative's office lobbied the EU aggressively, to put a stop to the WEEE and RoHS directives before they were passed. Thus, in an increasingly common role, the United States worked to weaken another group of nations' laws so as to benefit private corporations based in the United States. In fact, the U.S. threatened to take the WEEE issue up with the World Trade Organization (WTO) if certain provisions were not omitted.[115]

What is perhaps even more troubling is that even the WEEE directive, while arguably the strongest such legislation anywhere, will not address the *roots* of the e-waste problem. The treadmill-of-production theory emphasizes the relationship between growth in markets and socioenvironmental harm, and a critical examination of recycling concludes that this practice will actually enable that process. Unless growth itself is challenged, the social and environmental ills that activists bemoan will worsen. Gary Griffiths, environmental manager for a computer refurbisher in Britain, noted, "Given that the amount of WEEE [waste from electronics goods] is set to double by 2010, this means that the same amount now being disposed of to landfill and incineration may continue." A perverse effect of the WEEE directive is that as more material is collected for recycling, it may create a greater demand to export e-waste illegally to southern economies for dirty recycling. This is exactly what happened with bans on e-waste in landfills—intended to encourage domestic e-waste recycling—in the United States.[116]

European nations have signed the Basel Ban on toxic waste exports, but there are doubts about enforcement. Since Europe agreed to stop exports, the BAN team has been back to China and reports that although most e-waste comes from the United States, it is still "flowing out of Europe."[117] BAN's Jim Puckett declared, "So much harm has come under the green passport of recycling. Whenever someone says that word, it has the effect of making people swoon and think that everything is going to be lovely."[118] Thus, as we have already seen, ecological modernization in the global North may be occurring hand in hand with environmental inequality in the global South.[119]

Industry Responses

In addition to developing recycling and computer takeback programs, some electronics producers have gone further. Apple, Dell, H-P, Intel, and Sony have reduced or eliminated the use of polybrominated diphenyl ethers (PBDEs), a harmful fire-retardant chemical that is a rapidly expanding presence in the global ecosystem. PBDEs are called the "son of PCBs" because their chemical structure is similar to polychlorinated biphenyls. PCBs, the cancer-causing chemicals used for decades to insulate electrical equipment, were banned in 1977, but they were contained in so many products and are so long-lasting that they continue to affect the global food chain. Both PCBs and PBDEs are compounds that accumulate in the fatty tissues of humans and other animals. PBDEs are suspected of causing developmental abnormalities in children's brains.[120] The state of California and the EU have passed laws banning PBDEs, which are contained in electronic devices such as stereos and computers, foam cushions in furniture, car seats, carpet padding, and fabrics.[121] Lawmakers in Washington State and Maine are considering a ban, and momentum is building for a federal prohibition as well.

And although Sony's efforts to reduce the use of PBDEs are commendable, the corporation has also engaged in less than savory behaviors on other fronts. One particular action recently sent a chill through the environmental activist community. A leaked document suggested that the Sony Corporation had been monitoring—through surveillance—environmental activists who were participating in international campaigns to hold electronics manufacturers responsible for their toxic waste. An internal Sony document leaked to the InterPress Service outlined a presentation titled "NGO Strategy" made in July 2000 by Sony representatives at the European Information and Communications Technology Industry Association conference on environmental policy in Brussels. According to a report by the newsletter *Inside EPA,* Sony planned on using the intelligence it would gather "to track and potentially cripple activists efforts on a global scale."[122] The leaked document includes the names of activist groups targeting the electronics industry, including Greenpeace, Friends of the Earth, the European Environment Bureau, and the Silicon Valley Toxics Coalition. Sony described the groups as "highly active" and "well organized" with "global reach." SVTC's Ted Smith was startled to discover that the company was discussing his group's activities. He observed, "It seems that industry has spent an inordinate amount of time fighting the tide instead of doing what they need to do to clean up the industry."[123] Mark Small, vice president of environment and health and safety department with Sony in the United States, acknowledged that Sony was tracking the activities of environmental groups. He remarked, "We are obviously concerned about our image, and we want to make sure that if Greenpeace is pushing something we want to be on top of it."[124]

While certain corporations phase out harmful chemicals or work against environmentalist opposition, others continue to challenge the Basel Ban's restriction on the trade in waste officially destined for recycling because they argue this prevents legitimate recyclers from engaging in commerce.[125] In the next section, I present a framework that might be able to move stakeholders toward a more sensible and sustainable way of doing global business.

Concrete Solutions: An International Environmental Justice Framework

Leslie Byster, an activist with SVTC and the ICRT, states, "Each new generation of technical improvements in electronic products should include parallel and proportional improvements in environmental, health, and safety as well as social justice attributes. The mantra of the industry is 'faster, smaller, cheaper.' Our mantra is 'cleaner, greener, safer.'"[126] This new paradigm would constitute a direct challenge to the general structure and function of the global electronics industry with regard to health and environmental justice.

This vision of sustainability must be built on specific goals and steps. The Computer TakeBack Campaign proposed such a framework in its demands to the electronics industry. These demands are contained in the CTBC's "Take It Back Principles" (box 6.1).[127]

Challenging the World Trade Organization (and all free trade agreements) and those who would use it to facilitate environmental injustices and trade harmonization, environmental activists from more than one hundred NGOs recently declared in a signed public letter, "We must level environmental standards up, not down. Trade associations must not be allowed to dictate environmental health policy."[128] The CTBC worked with electronics producers and recyclers to set an example to those firms that

Box 6.1
The Computer TakeBack Campaign's Take It Back Principles

Take it Back

• Financial and/or physical responsibility. Manufacturers and distributors of electronic goods must take financial and/or physical responsibility for their products throughout the entire product lifecycle, including take-back and end-of-life management.
• Infrastructure development. When firms engage in extended producer responsibility this will develop an environmentally sustainable infrastructure for the collection, re-use, remanufacturing, and recycling of electronic equipment.
• Stop hazardous waste exports. The U.S. federal government should ban exports of hazardous materials contained within electronic waste equipment.
• Taxpayer relief. Corporations—not governments and taxpayers—should pay for electronic waste collection, recycling, and disposal.
• Community re-investment. The recycling infrastructure that emerges from a comprehensive "take back" system should support local forms of economic development associated with recycling, re-use, and remanufacturing.
• Internalize costs. Electronics producers should internalize the costs of environmentally progressive production, rather than shifting this burden to consumers. This should also prevent producers from being placed at a competitive disadvantage.
• Recycling goals. The electronics industry should meet stringent recycling goals.

Make it Clean

• Adopt the Precautionary Principle. Where there is a threat to human health or the environment, a precautionary approach requires taking preventive action even before there is conclusive scientific evidence that harm is occurring. The federal government should develop and implement strict protocols for testing chemicals and mixtures before they are introduced into the markets. This regulatory approach departs from the traditional paradigm in that it shifts the burden of proving that a chemical is safe from the state to the producers of that substance. The City of San Francisco has formally adopted the Precautionary Principle, but a national standard is needed.
• Phase-out hazardous materials. The electronics industry should cease the use of materials that are harmful to human health and the ecosystem (for example, lead, mercury, cadmium, brominated flame retardants, chlorinated solvents, etc.).
• Proper handling of hazardous materials. Until safer substitutes are available, manufacturers of electronics should do their best to protect their workers, the public, and the environment from hazardous substances used in the industry.

Box 6.1
(continued)

• Design for the environment. Manufacturers should design products for greater durability, longevity, upgradability, and eventual disassembly. In addition to using less toxics in the production process, every effort should be made to reduce the use of water, energy, metals, plastics, and other natural resources throughout the product lifecycle.
• Closed-loop recycling. The electronics industry should design products with high recycled content and with remanufactured components in new products where possible. Closed material cycles are to be encouraged.

Zero Waste

• The goal is to restrict (ban) all discarded materials from landfills and incinerators and to encourage environmentally sound recycling practices.

Fair Labor

• Protect workers. The electronics industry should end exploitative and unsafe labor practices throughout the product chain, including hazardous working conditions and the use of prison labor.
• Fair pay. The industry should institute a living wage for its workers throughout the product chain, including for employees of subcontractors.
• The right to organize. The industry must recognize and respect its workers' legal right to organize in unions and to engage in collective bargaining, throughout the product chain.

Source: http://www.computertakeback.com

might doubt the feasibility of EPR. David Wood, former codirector of the CTBC and former executive director of the GrassRoots Recycling Network, stated "What we hope to do is showcase companies [that have good recycling practices] like Scientific Recycling, apply pressures throughout the electronics recycling industry and pull performance upward."[129] In February 2003, sixteen electronics recycling firms announced that they had signed a pledge to uphold stricter standards for processing electronic waste, including discarded computers, cell phones, televisions, and monitors that contain hazardous waste.[130] The "Electronic Recycler's Pledge of True Stewardship," authored by the CTBC, SVTC, and BAN, was a direct outgrowth of the global outcry over e-waste dumping in southern nations. This is strong evidence that these social movement organizations are articulating a political economic opportunity structure framework that links the state and corporations as the targets of change in a global context, in the hope that such policy changes will disrupt or even reverse the production of environmental inequalities across race, class, and national borders.[131] It can also be said that electronics takeback systems constitute the environmental justice movement's success at institutionalizing a return-to-sender policy among corporations and governments. While activists fighting other forms of transnational waste trading and dumping (such as municipal and agricultural wastes) have had to literally send the waste back to their nation or region of origin, it appears that e-waste activists have adopted another tactic: design for environment and take-back systems that reduce toxics at the point of production and recycle them at the end of life.

Conclusion

Consistent with the political economic opportunity structure framework, e-waste activists have not limited themselves to a state-centric approach, with regard to movement targets. In addition to pushing states to pass laws that ban e-waste from landfills and set up recycling programs, they have also successfully pressured the electronics industry and specific companies to change their practices. A political economic opportunity structure approach that targets states and corporations responsible for pollution is the most productive strategy for achieving local and global environmental justice. Environmental activists in the North and South have also clearly framed the problem of electronic waste dumping as one that produces class, racial, and national hierarchies in environmental protection, thus constituting environmental injustices. These inequalities are exacerbated in some cases by the desire to embrace electronics and high-technology development among Asian nation-states. Thus, activists must work across race, class, and national differences to produce progressive political, economic, and social change for environmental justice.

All of this comes back to the importance of mobilizing local, national, and transnational movement networks and using them for the effective production and exchange of ideas, information, data, and images and for pressuring states and corporations to change their policies. For activist networks organizing around the high-tech industry, there is a tension and irony in that they use the very technologies whose toxicity they confront. Activists are also clear about the limits of high technology in political organizing campaigns, where trust must be established across long distances. Ted Smith notes:

> We use the Internet a fair amount, but there is no substitute for meeting people and developing the kind of trust you need for working internationally. Without that, the Internet is so impersonal and cold a communication media that it's not very effective. If I don't know the people on an email listserv, then I'm going to be a whole lot less willing or able to engage in a serious or strategic conversation, if I don't know them personally. I think that's true more internationally than locally or regionally because the gaps are really huge. Our visit to Taiwan was very important because it was an opportunity to learn firsthand about what was actually going on with people. It was our chance to make personal contact with an incredible array of people. They are trying to get us to come back this spring and we're trying to get them to come back here, including a group of injured workers. It certainly helps that there are people here in the Bay area that are part of the TEAN network [Taiwan Environmental Action Network]. The ability to maintain close personal contact is key.[132]

Smith speaks to the need for, and the importance of, face-to-face interactions for the development of global organizing campaigns. In the shadow of some of the world's most powerful institutions, the electronics industry and its collaborating nation states, Smith and others involved in many environmental justice activist networks articulate a vision of hope while creating a robust global public sphere.

7

Theorizing Global Environmental Inequality and Global Social Movements for Human Rights and Environmental Justice

Studies of the transnational waste trade have paid far more attention to the problem of hazardous exports than opposition to it. In this book, I have addressed this gap in the literature by exploring some of the most critical moments and important dimensions of transnational environmental justice movement networks organizing around global environmental inequalities. I have also addressed the gaps in social movement theory, which has not adequately theorized race or the role of capital in creating structures of oppression and targets of resistance, and race theory, which has paid insufficient attention to the role of environmental inequality or corporate power.

In late modern nation-state governance and transnational corporate practices, we witness a fundamental transformation in the relationship between these institutions and the environment—an exponential increase in the production and use of hazardous chemical substances and an intensification of social hierarchies within and between societies. What this means is that the very idea of the modern nation-state in a global economy is associated with invidious social divisions and the existence of toxins that permeate every institution, the human body, and the natural environment. Industrial production, communication, transportation, agricultural systems, and militarization are all normal functions of nation-states, and all produce environmental inequalities as a matter of course, not as a result of market failures or aberrations. Not surprisingly, then, the late modern nation-state and corporation are spaces where we observe struggles over these persistent social hierarchies and environmental harm. In this book, I have integrated theories of modernity, race, and class with theories of environmental conflict and social movements. I developed a framework that centers race, class, and the environment in debates concerning globalization, and I underscore that transnational social change efforts by grassroots nongovernmental organizations (NGOs) are not only possible but are necessary and sometimes quite successful.

New Directions for Theory

Drawing on theories of the society-environment interface, I contend that if ecological modernization is at work in the world, it is largely in those national and regional spaces privileged by whiteness or wealth, which enjoy these benefits at the expense of nations, regions, and spaces where poor populations and people of color struggle with weak economies. That is, environmental justice studies allow us to both confirm and challenge ecological modernization, because industrial risks adhere to the path of least resistance. I also find that the risk society and treadmill-of-production theories are useful for explaining the production and maintenance of global environmental inequality. Both theories underscore the power of nation-states and capital to shape society and the very landscape on which it rests. Both theories advance useful frameworks for analyzing class conflict in the context of environmental politics. But since neither theory examines the role of race and racism in the ways in which nature is imagined, produced, and manipulated, I link these models to research from race theory.

Borrowing from sociology, critical race studies, and philosophy, I link theories of race to theories of environmental change and class struggle. Guinier and Torres's work connects well with Beck and Schnaiberg in ways that allow us to theorize both the toxicity of racism and class domination, and the racism and class hierarchies that emerge from and are reinforced through toxic production and disposal regimes. That is, both race and class inequalities and chemical toxins operate and cooperate in ways that cause harm across social, spatial, and temporal boundaries, while also producing opportunities for creative resistance among communities that might normally be quiescent and disconnected. Numerous case studies presented in the preceding chapters demonstrate these dynamics. Consider the examples of pesticides in the Bahamas and Mozambique. Both cases were the result of international environmental racism and inequality and revealed that toxics operate very much like racism and class domination:

- Obsolete pesticides in the Bahamas and Mozambique posed risks to both humans and the environment.
- The toxins shaped the earth and its landscape literally and figuratively, as they were used to produce agricultural products and as the discovery of the stockpiles transformed people's imaginations about the local environments, as more fragile and vulnerable.
- The pesticides were simultaneously invisible and blatant. That is, activists observed barrels of waste, but the chemicals contained within them were odorless and undetectable by other human senses.
- The pesticides arrived in both locations because they migrated through space, over political, cultural, and national borders.
- The toxics migrated through time, across nearly a generation in the Bahamas case.
- They provided privatized benefits to certain people, institutions and organizations, and nations, while imposing and publicizing the costs on others.
- The pesticides were mobilized as powerful frames for organizing resistance movements and bringing people together across social and spatial boundaries.
- As a result of both campaigns' success in advancing a return-to-sender (RTS) strategy, the toxins operated like a boomerang and circled back to have an impact on members of far-away societies.

Circularity, Boomerang Effects, and the End of the Other?

One of the themes that connects all of these problems is their circularity. That is, racism, class domination, toxics, and social movements intersect and collide to reveal how the costs of environmental injustice are borne by everyone, ushering in what Beck called the "end of the other," a democratization of risk.[1] The circularity of toxics, racism, class inequality, and social movement power is illustrated well in the metaphor of the boomerang. A boomerang starts at the point of ejection and, if thrown properly, ultimately returns to that same point. It more or less follows a circular path. When the boomerang hits its target, the path is interrupted, and when that happens, it does violence, or at least has an impact. Toxics, racism and class domination, and social movements each create boomerang effects, separately and in combination.

Toxics

Through the circle of poison, toxics travel from their point of origin throughout ecosystems and human bodies, producing illness, death, and disruption across space and time. Today toxic chemicals are so powerful that they put at risk not only those immediately exposed to them, but also persons living far away from points of production, use, and disposal. For example, indigenous peoples in the arctic North find persistent organic pollutants in their bodies, hundreds, if not thousands, of miles away from the sites of production. This is environmental injustice, but it is also reflective of how everyone to some degree suffers from the toxicity of late modern capitalism and nation building.

Race and Class Exploitation

Racism and class domination operate like boomerangs as well, originating from persons, groups, institutions, and ideas and harming others targeted for exclusion, devaluation, persecution, and exploitation. And although the targets of racism and class domination suffer the greatest from this social poison, it eventually circulates back to those who initially benefit from it. That is, social inequality and exploitation cost societies in the form of economic losses, psychological trauma, and social discord and upheaval. Scholars and antiracist activists have long argued that racism ultimately hurts everyone.[2] Feagin, Vera, and Batur write, "Americans should see white racism for what it actually is: a tremendously wasteful set of practices, legitimated by deeply embedded myths, that deprives its victims, its perpetrators, and U.S. society as a whole of much valuable human talent and energy and many social, economic, and political resources."[3] Feagin, Vera, and Batur argue that white racism has become a ritual of societal waste and collective destruction.[4] Entire societies, and all racial groups and classes, pay a heavy price as a result.

Social Movements

Transnational social movements produce boomerang effects as well. When local governments refuse to heed calls for change, transnational activist networks create pressure that "curves around local state indifference and repression to put foreign pressure on local policy elites. Thus international contacts amplify voices to which domestic governments are deaf, while the local work of target country activists legitimizes efforts of activists abroad."[5] It is the interaction between repressive domestic political structures and relatively open structures in other nations that produces this boomerang. And as we have seen in the case of the RTS tactic that transnational environmental justice networks employ, this boomerang effect takes a curious turn. That is, countering the traditional view of transnational movement networks fighting corrupt global South regimes and seeking assistance from activists in presumably more democratic northern nations, frequently environmental justice networks find that global North governments are just as closed and corrupt as southern states and that the boomerang effect must be focused on those wealthier states. The RTS strategy is the ultimate example of circularity and the boomerang effect because it involves ordinary people working through transnational activist networks battling toxics and social injustice by returning these poisons to their point of origin, thus aggressively making the collective costs of environmental inequality known to all. In other words, while the boomerang effects of racism, class exploitation, and chemical toxins ultimately also harm their producers, transnational social movements serve as a kind of insurance policy that returns these poisons to their senders, just in case ecosystems and social consciences decide to move "with all deliberate speed."[6]

In everyday parlance, the boomerang effect speaks to the adage, "what goes around comes around." Or, as environmental justice activist Orrin Williams is fond of reminding white suburbanites, "no one is exempt" from the reach of environmental toxins.[7] Although I agree with this sentiment and view circularity as a critical concept that links all peoples through ecosystems and social systems, persistent social inequalities suggest that this does not yet constitute Beck's "end of the other."[8]

Wither the State?

Many scholars have raised the question as to whether nation-states are even capable of addressing global social and environmental problems, largely because of their limited geographic reach.[9] This is a valid concern.

But what I am suggesting is that this is a problem primarily because of the more formidable power of TNCs and *their* extensive geographic reach. The problem of waning nation-state power has been an issue for some time, as corporate power has grown immensely since the 1940s.

Social movement organizations confront local, regional, national, transnational, and global political processes in their campaigns. In the case of battles over e-waste, activists face political processes at the state, national, transnational (an example is the European Union and its Directive on Waste Electrical and Electronic Equipment and Directive on Restrictions of Hazardous Substances), and global scales (for example, the U.S. plans on using the World Trade Organization to challenge extended producer responsibility laws and the existence of the Basel Ban).

Leading social movement scholar Sidney Tarrow contests the notion that states are receding in their relevance because modern states originally developed in a dialogue with social movements and today have created transnational strategies and organizations working for their interests.[10] George Lipsitz also rejects the idea that nation-states matter less in a global economy. Although it may appear that the power of transnational corporations (TNCs) supersedes the nation-state, he argues that the state continues to be an indispensable component of the global system. The state serves as a crucial resource for TNCs by supplying mechanisms for capital accumulation and technological innovation through direct investment by governments, indirect support for research and development, tax abatements, and R&D spending on the infrastructure of global capital. For example, high technology and the Internet, which we now associate with the entrepreneurial spirit of business leaders and large firms, were originally developed by the U.S. military. The state also supplies TNCs with political regulation through direct repression of insurrections and strikes and by ratifying free trade agreements, which deprive citizens of the power to use politics to challenge corporate domination, environmental pollution, labor exploitation, and business monopolies. The state also disciplines labor forces and imprisons or ignores surplus labor. The elite classes and racial/ethnic groups in nation-states from Venezuela to Malaysia benefit from this dynamic. The very existence of nation-states discourages internal hostilities among domestic populations and refocuses resentment against external threats rather than those elites at home.[11] Consider the current moment in the United States: social inequality between social classes and racial groups is rising, the labor movement is weak, environmental protection is seriously threatened, and the rights of citizens are being stripped away while an inordinate degree of the political focus is on terrorist threats and rogue regimes abroad.[12] As I have argued earlier, rather than receding, many states have simply shifted their focus away from public welfare approaches toward the support of privatization and a broader neoliberal agenda, thus adopting a corporate form. That orientation reveals the influence and reach of the ideology of privatization and therefore serves to strengthen the position of TNCs. Hence, the power of TNCs is considerable and must be seriously considered and theorized in relation to social movements.

The scholarly literature on transnational social movements and global civil society is marked by the absence of a serious consideration of political economy in opportunity structure models. This would seem important if we are considering the role of TNC activities because they constrain the power of states and shape the political environment in which movements for environmental justice operate.

While race theorists argue that nation-states constitute "racial states" because they reproduce racial logics, the corporation as a racial institution is an idea that scholars might also pursue and develop.[13] Sociologists studying racial formation in the twentieth and twenty-first centuries have argued that the post–World War II period signaled a historic break in the terrain of race relations in the United States and globally. During this time, social movements and world opinion challenged overt legalized racism, which produced civil rights gains for people of color in the United States and pushed blatant racist practices to the realm of the institutional and the less visible. Hence, the rise of the "postracial" or "color-blind" discourse that pervades legal and political considerations of race today. Thus, the discourse about racism is color blind, while institutional practices and policies continue to reinforce racial hierarchies.[14] What is truly troubling, however, is the number of scholars who themselves have uncritically adopted the "postracial" paradigm embraced by politicians and media pundits without understanding its construction as a tool to obscure persistent racial hierarchies.[15] No less troubling is the fact that many scholars studying social inequality have also found it less acceptable to focus on class divides in society; thus, Marxist—or, for that matter, virtually any class-based research—approaches have fallen out of favor as "irrelevant," precisely at a time when class inequalities have reached historic records.[16]

In essence, the absence of a discussion of class politics is a "class-blind" approach that parallels the "color-blind" orientation, and both are pervasive in politics and media coverage as well.

What the historic break argument and the color-blind and class-blind paradigms ignore is that at the same time that era began, after World War II, we saw the rise of transnational corporate power. That event revealed another break wherein nation-states no longer enjoyed monopolies in the arenas of governance, in general, and in defining and producing racial and class hierarchies, in particular. Today TNCs wield as much power over producing and guiding racial and class politics and the material landscapes on which these relations occur as do nation-states. We live in a world where nation-states are not the only entities that can colonize others and build empires. Corporate colonization is just as pervasive, as numerous corporate polities are active around the globe, through which these private organizations exercise as much political, economic, and cultural influence as governments do.[17] TNCs contribute to, produce, and benefit from racial and class inequalities. The TNC is the most powerful institutional force today.[18] There is no better evidence of this than the problem of global environmental inequality. Environmental inequality in this late modern era is intimately linked to the rise of TNCs, because during that same moment when the historic break in race relations occurred and TNCs emerged with greater force than ever before, some of the most influential of these corporations were engaged in the manufacture and sale of toxic chemicals. Chemicals became an integral component of production across most major industries and became embedded in the very fabric of industry, society, human bodies, and ecosystems. These chemicals continue to have their greatest negative impact on the global South: people of color, the working classes, indigenous peoples, poor persons, women, children, immigrants, and economically struggling nations.

The Future of Environmental Justice Movements: The United States and the Rest of Us

In the past two decades, global environmental problems have entered the public's consciousness and the agendas of environmental organizations around the world. For the U.S. environmental justice movement, this represents a unique challenge owing to the privilege of living in a global North nation (although inhabiting the South of the North) and accountability for contributing to the globalization of environmental harm. Cities in the global North are the point of origin for much of the world's toxic wastes, and even environmental justice groups representing peoples in global North nations enjoy enormous privileges relative to people's organizations in global South nations.

One thing that may strike the reader from the cases in chapters 4, 5, and 6 is that despite the common view of global South nations as desperately poor, environmental justice activists in many of these countries have often been able to exert considerable power over TNCs and nation-states, often more so than activists in northern nations. For example, it is rare to hear a report that a poor neighborhood or community of color in the United States or United Kingdom successfully returned a waste shipment to the sender. This would be a remarkable display of power and would be celebrated as a shining example of environmental justice. Yet activists in the global South have repeatedly achieved this very outcome, this critical indicator of environmental justice. Thus, while the cynic might be inclined to conclude that these international environmental justice collaborations are little more than projects that reinforce the global geopolitical hierarchies among activists and NGOs in the North and South (and certainly there is some truth to that), they also reveal that these struggles frequently exert significant leverage and enjoy successes that environmental justice groups in the global North rarely see. Of course, one might argue that successes in the South are frequently due to the exertion of external influence over weak states. There is merit to that point as well. However, it is also true that many governments in the South have historically had little patience for civil society pressures. Moreover, in many cases, it is the global North governments and corporations that are targeted and eventually succumb to the force of international activist pressure, and few could argue that these are weak states or fragile institutions.

So perhaps what Heeten Kalan, a South African environmental justice activist, recently declared is true: "Despite the fact that the environmental justice movement, or rather, the *language* of environmental justice, began in the United States, some of the most exciting and innovative action going on in terms of environmental justice organizing is occurring *outside* of the United States."[19] The U.S. environmental justice movement could learn valuable lessons by paying more attention to the movement in the global South. Specifically, groups in the global South are very clear that they *must* reach out and network with allies of different ethnicities, races, and nationalities around the world. This is something some U.S. environmental justice activists have been hesitant to do.

Many U.S.-based environmental justice activists and nongovernmental organizations have resisted this approach and have insisted on working mainly through networks of people of color. The tensions between Anglo and people-of-color activists in the United States are often high and have long-running, deep histories that involve pain and betrayal, so this is understandable.[20] However, for the U.S. environmental justice movement to achieve long-term transformative social change, it will have to branch out beyond the people-of-color networks. It is not as if global South activists and their communities have no memories of the experience of colonialism, racism, and exploitation at the hands of other nations and ethnic groups. Yet many of them have made the pragmatic decision to join forces with allies across borders to increase their leverage at home and elevate the visibility of their struggles beyond their domestic national spheres. Heeten Kalan does this kind of work in the United States and Africa and argues

> A major lesson for me is that the broad coalitions that we seek to build in movements aren't just tactical; they're important, and they have to happen because our issues are so interconnected and interlinked. . . . I think part of it is that South Africans tend not to approach our political activism from a point of identity politics. And so for me, all these folks that I decide to work with in the U.S., I see them as allies. They're our allies, they're people who can help our struggles, they strengthen the work. I know for a fact that there are white folks in the U.S. South, white folks in Maine, white folks in New Hampshire, who are poor as hell and are getting screwed by big multinational companies. Now are those issues not important to me? Absolutely they are, and they should be because it's the same bloody company that's killing people in South Africa. And if I can't make the link, then I've closed myself off to this work. I am closing myself off to see the potential that we can actually gain and the strength we can actually gain by all of us working together. Now, having said that, I have my struggles with predominantly white organizations around issues of race, but I'm not willing to throw the baby out with the bathwater, and I'll engage them on those issues, and I'll be direct with them on those issues, and I'll work with them in pushing for a world that's cleaner and safer for all of us. They bring good resources, they bring good knowledge, good experience and they bring their passion to this.[21]

If U.S. environmental justice movement networks are going to work more aggressively on a transnational scale, they must consider these questions. Building a transnational network can have clear benefits. Livaningo activist Anabela Lemos (Mozambique) explained that having such a network

> helps a lot because sometimes when you are fighting and we are very close to the issue, then we can't see any more than what is in front of us. If you have a network and an issue, you can say "this happened, what should we do?" Not that you're automatically going to do what they tell you to do; it is just advice and you see a different way of reading things . . . because realities are different in different countries. Something that you can apply in your country doesn't apply to us, so we have to find our own way, but with support and information.[22]

Many African environmental justice activists like Kalan and Lemos are therefore collaborating across social, spatial, and cultural borders to increase their leverage and build the movement. Indigenous organizing is also ongoing within Africa among Africans, as evidenced by the collaborations among groups like Livaningo, the Environmental Justice Networking Forum in South Africa, and NGOs in Nigeria, Cameroon, Zimbabwe, and elsewhere. Anabela Lemos made it clear that African environmental justice NGOs see the critical value of working together among themselves: "And what we are trying to do is more African-based networking, having as much resources as possible between us. So don't get the idea that we are going always from the third world country to the first world country, so that's why we are trying now to have networks in Africa itself to support each other . . . in the civil society."[23]

It is important to understand that the cases in this book involve virtually no participation by the "big ten" mainstream environmental organizations in the United States, such as the Sierra Club, the Natural Resources Defense Council, or Environmental Defense, although these groups do engage in international campaigns. The environmental groups considered in this study are explicitly international in their reach and much more politically progressive, if not radical, organizations and networks when compared to most of the mainstream U.S. groups.

What is also noteworthy is the absence of U.S.-based environmental justice groups from communities of color in many of these transnational environmental justice coalitions. Although U.S. environmental justice groups have become more active in international networks in recent years, these kinds of actions remain relatively rare. What transnational movement networks have necessarily learned, which local and national networks in the United States are still learning, is that what I term the hyperspatiality of risk dictates that environmental harm will follow the path of least resistance from one location to the next and that this is their responsibility. So when transnational environmental justice activists succeed in defending one community from the scourge of environmental injustice, this is just the first of many steps in the process of seeking environmental justice, when for local and national movements it is often the only goal.[24] Transnational environmental justice movement networks are acutely aware of the ways in which risk travels across social and geographic borders, because these networks were born out of the recognition of this fact. This attention to the hyperspatiality of risk is something on which local and national environmental justice movements must place greater emphasis.[25]

Many environmental justice activists in the United States, while relatively privileged economically, still exist in the "South of the North," and must therefore view their struggle as an integral component in the global environmental justice movement. Then, and only then, will the U.S. environmental justice movement truly move forward and become the force that it has always had the potential to become.[26] The global movement for environmental justice is growing and stands to gain much from the participation of U.S. activists and organizations in communities of color and all working-class communities.

International Human Rights, Environmental Justice, and Global Citizenship

The Emergence and Growth of Human Rights Law and Conventions

Civil rights are distinct from human rights in that the former are generally guaranteed only in the domestic sphere, while the latter are intended to be global, universal norms by which all nation-states must abide. The first formal global human rights legal framework was set in motion in 1948 with the United Nations General Assembly's approval of the Universal Declaration of Human Rights. The development and evolution of human rights norms and law has special significance for struggles against environmental inequality because the rights that many environmental justice activists seek to protect are actually considered, if not guaranteed, in many human rights legal frameworks. For example, the UN's Draft Declaration of Principles on Human Rights contains the universal human right to a "secure, healthy, ecologically sound environment."[27] The 1992 Rio Declaration, which came out of the UN Conference on Environment and Development, or the "Earth Summit," declared the right to a clean and healthy environment as a fundamental and inalienable human right. That declaration was also made possible by the work of numerous environmental and human rights activists working with the UN and various nation-state representatives. A U.S. judge recently referred to that declaration in the case of indigenous peoples in Ecuador who sought the right to sue the Texaco oil corporation for environmental harm to their land. That ruling also made it clear that corporations, not just states, were to be held accountable for human rights violations. The struggle for racial justice is also deeply acknowledged and present in international human rights law and conventions. The Universal Declaration of Human Rights explicitly proclaims that all human beings are entitled to freedom, dignity, and rights without regard to race, color, or national origin. More specifically, the UN approved the International Convention on the Elimination of All forms of Racial Discrimination in 1965, which rejects all practices, policies, and discourses among nation-states that support or produce racial discrimination. The 1979 UN Convention on the Elimination of All Forms of Discrimination against Women offered similar language directed at gendered violence and patriarchy. And since the UN has also passed resolutions condemning colonialism in all forms, the spaces created by these global rules have become critical sites of political mobilization by environmentalists, racial justice activists, feminist and women's movements, and advocates for corporate accountability.[28]

The Environmental Justice Movement and Human Rights

The environmental justice movement is steadily moving in the direction of linking environmental justice discourses and actions to human rights frameworks. Monique Harden and Nathalie Walker are environmental justice and human rights attorneys who cofounded and codirect the group Advocates for Environmental Human Rights (AEHR), based in New Orleans, Louisiana. AEHR is a "public interest law firm whose mission is to defend and advance the human right to a healthy environment."[29] Toward that end, AEHR recently petitioned the Organization of American States, charging the U.S. government with violating the human rights of African American residents of Mossville, Louisiana, since they suffer from environmental racism, and the environmental regulatory system provides little protection against this practice. AEHR's work represents a new and productive direction in environmental justice movement organizing in the United States and globally. Connecting environmental justice and human rights elevates and deepens the discourse, the struggle, and the framework within which activists' claims can be made and resolved. A human rights framework internationalizes the movement's concerns in ways that have rarely been articulated, which can connect local activists and campaigns to a much broader group of actors, discourses, policies, and possibilities.

The environmental justice movement was recently ignited in Central and Eastern Europe (CEE) when the Coalition for Environmental Justice (CEJ) formed there in 2002. The CEJ is made up of scholars, activists, and lawyers from environmental and human rights organizations in Bulgaria, Czech Republic, Hungary, Macedonia, Romania, and Slovakia. In 2005, coalition members decided that they would benefit from an exchange with U.S. environmental justice scholars, activists, and lawyers. The ultimate goal was to launch an effective transnational initiative to promote environmental justice on both sides of the Atlantic. So in October 2005, several dozen activists, scholars, and attorneys (I was among them) from the United States and CEE nations convened a meeting in Budapest, Hungary, to launch the Transatlantic Initiative on Environmental Justice, to unite activist and research campaigns and environmental justice and human rights initiatives from both parts of the world.

One of the focal populations for this environmental justice and human rights work in the CEE are the many Roma communities in the region that are the targets of daily abuse at the hands of states, majority ethnic groups, and individuals. Examples of contemporary anti-Roma abuse include sterilization programs for women of childbearing age; year-round forced evictions; lynching; daily racial profiling and police beatings; random police raids on homes; brazenly anti-Roma public campaigns by political parties; refusal to enroll Romani children in mainstream schools, forcing them to go to schools for the mentally retarded; denied access to health care; and housing segregation.[30] This is vicious, blatant racism that would remind most U.S. residents of Mississippi in the 1930s.

Not surprisingly, Roma populations in CEE also face a host of environmental injustices linked to racism, poverty, and housing segregation, such as exposure to flooding and toxic materials from mines and waste dumps and a lack of basic services. One civil society group working in Roma communities in CEE is the European Roma Rights Center (ERRC). The ERRC has worked on some of the worst cases of environmental racism directed at Roma people in Europe. Larry Olomoofe, a human rights trainer for the ERRC stated, "If you put your finger anywhere on a map of Europe where Roma are located, you'll find environmental problems."[31]

There are two specific outcomes of the Transatlantic Initiative that are noteworthy. One is a collaboration among several groups within the coalition, including university students and scholars and Romani activists and environmentalists, who are challenging the traditionally racist view that Romani people prefer to live near garbage dumps. In Sofia, Bulgaria, they are documenting the fact that Roma are often forced by cities and by threat of public violence to live in close proximity to garbage dumps, a classic example of environmental racism. The coalition is also documenting the fact that Roma who are taking objects out of the dumps for reuse or resale are what, in any other context, we might call recyclers or resource recovery workers. These individuals contribute to Bulgaria's national recycling efforts. Thus, this coalition is reframing the discourse on the work Roma are doing in garbage dumps and on why many Roma live in close proximity to these dumps in the first place.

The second outcome of the Transatlantic Initiative is another collaboration. Global Response and the ERRC teamed up to launch an international letter-writing campaign to persuade the UN to relocate Roma families from a toxic waste site in Kosovo. In 1999, at the end of the Kosovo war, ethnic Albanians (who themselves face a host of oppressive circumstances in Kosovo at the hands of Serbia) violently expelled approximately four-fifths of Kosovo's Romani population from their homes. Romani people were kidnapped, abused, murdered, and raped. Whole Romani settlements were burned to the ground, and North Atlantic Treaty Organization forces did not interfere to stop this ethnic violence. The UN eventually relocated more than five hundred Roma from the area to northern Kosovo. Unfortunately, they were placed in refugee camps where they were exposed to severe lead poisoning because the sites were located near a major mine and lead smelter. The World Health Organization (WHO) conducted a study in these camps and found that 88 percent of the children under age six had blood lead levels in the highest category, described as an "acute medical emergency."[32] In 2005, the WHO declared that the camps should be evacuated immediately, calling the situation "the worst environmental disaster for children in the whole of Europe."[33] The International Committee of the Red Cross also demanded that the camps be evacuated. Why, after several years, were more than five hundred people still living on lead-contaminated soils and breathing lead-contaminated dust? The ERRC argued that there was only one explanation: racism. Racist sentiment against the Roma is so pervasive that the UN was able to keep them in the camps for six years without risking a major public outcry. So the ERRC and Global Response launched a letter-writing campaign to "call this injustice by its name: environmental racism" and urged their members and the public to join Roma community activists in their demand that the UN immediately relocate the Roma refugees and provide the medical care they need.[34] After seven months, nearly all Roma families were safely relocated.[35]

The Transatlantic Initiative on Environmental Justice is an example of how environmental justice and human rights frameworks and movements are merging and are being articulated by global activist networks, even in regions like Central and Eastern Europe, which is often excluded from discussions and actions focused on global justice.

Remembering and Recovering Human Rights Discourses

If human rights represents a newly visible direction for environmental justice movements, we must remember that it has been present from the movement's earliest days in the United States. Consider the Principles of Environmental Justice, adopted at the First National People of Color Environmental Leadership Summit in 1991 (see the appendix at the end of the book). Of these seventeen principles, at least seven speak explicitly or implicitly to human rights and in reference to various UN conventions that guarantee these rights. For example, the first principle calls for the "right to be free from ecological destruction"; the second calls for the right to be "free from any form of discrimination or bias"; the fourth invokes the "fundamental right to clean air, land, water, and food"; the fifth affirms the "fundamental right to political, economic, cultural and environmental self-determination of all peoples"; the eighth cites the right "to a safe and healthy work environment"; principle ten argues that governmental acts of environmental injustice are violations of international law and of "the Universal Declaration On Human Rights, and the United Nations Convention on Genocide"; and principle thirteen calls for the strict enforcement of the "principles of informed consent."[36] Separately and together, these principles speak impressively to a body of international law and human rights that has been in development for six decades.

Environmental justice movement groups have always emphasized human rights, implicitly or explicitly. Consider the very names these groups have given themselves:

- People for Community Recovery—emphasizing compensation, if not reparations, for harms done to entire communities
- El Pueblo para Aire y Agua Limpio (People for Clean Air and Water)—calling attention to access to these two elements as basic rights
- The Southwest Network for Environmental and Economic Justice and the Colorado Peoples Environmental and Economic Network—underscoring the inseparable ties between environmental and economic rights
- Detroit Workers for Environmental Justice—linking the workplace and residential communities in people's desire to be free of toxic contamination in both spaces
- Gulf Coast Tenants Association and the Asian Pacific Environmental Network—groups defining safe and affordable housing as an environmental justice issue and as a fundamental human right.

The U.S. environmental justice movement has long advocated for environmental justice through language that speaks directly to broad human rights discourses and frameworks.

Conclusion

Environmental justice and human rights discourses and movements are slowly coming together as a global force. Articulating a vision of global environmental justice and human rights, GAIA activist Ann Leonard spoke of her network's hopes for the future:

GAIA has a vision of a just, toxic-free world, without incinerators. Greater globalization of companies and governments creates problems for public and community participation and governance. We have the *human right* to be in a clean, chemical free environment, down to the level of the individual body. How do we stop making all this waste and who makes these decisions? We want to shift the activist mindset from NIMBY (Not In My Back Yard) to NOPE (Not On Planet Earth).[37]

Echoing Leonard, activist-scholar Robert Bullard writes that the environmental justice framework "incorporates the principle of the right of all individuals to be protected from environmental degradation."[38]

These statements sum up what so many activists interviewed for this study say they work for. This is a vision of global environmental justice and human rights that involves creating sustainable and equitable communities and transforming corporate and state policies for the common good. These global justice activists acknowledge that the political economy is deeply imbued with race, class, gender, and national inequalities and that nation building in privileged spaces of the globe is made possible through the harm visited on other spaces, which are occupied by people who are devalued because of their cultural differences or relatively weak economies. They recognize that if ecological modernization is occurring anywhere in the world, it is not in the global South; in fact, ecological modernization in the North may be made possible by the transfer of the most hazardous wastes and technologies southward. These networks of activists target what they view as the sources of power and roots of the problem in the global political economy: states, corporations, international financial institutions, culturally insensitive environmental organizations, and the ideologies of racism and fiscal greed that place profit before people and the environment before people of color and the poor. This is the political economic opportunity structure in which they puncture holes and create points of access to interject their voices, bodies and actions, information and data, and imagination and hopes in order to create a more just and sustainable world.

All of this work indicates that transnational movements for environmental justice have become quite sophisticated at combating global environmental inequalities through the engagement of a range of institutions, thus developing an emerging form of global citizenship and, by extension, a burgeoning transnational public sphere.

Short of accomplishing such ambitious goals, what activists can achieve today are more just, equitable, and respectful collaborations among themselves. Sathinath Sarangi, one of the lead organizers fighting the Dow Chemical Corporation so that it will assume its responsibilities stemming from the 1984 chemical disaster in Bhopal, India, offers a hopeful reminder and vision of transnational collaboration: "I have seen some really good magic happen when I see people come from different communities to tell their stories. And so many practical solutions come up; practical ways of sharing things come when they talk. This is magic I have seen."[39]

It is my hope that in this book, I have offered a window into some of that magic that occurs when activists collaborate across racial, class, linguistic, cultural, and national borders to oppose the legacies and continuing practices of environmental injustice and articulate a vision of hope while creating a robust global public sphere.

Appendix
Principles of Environmental Justice

PREAMBLE

WE THE PEOPLE OF COLOR, gathered together at this multinational People of Color Environmental Leadership Summit, to begin to build a national and international movement of all peoples of color to fight the destruction and taking of our lands and communities, do hereby re-establish our spiritual interdependence to the sacredness of our Mother Earth; to respect and celebrate each of our cultures, languages and beliefs about the natural world and our roles in healing ourselves; to insure environmental justice; to promote economic alternatives which would contribute to the development of environmentally safe livelihoods; and, to secure our political, economic and cultural liberation that has been denied for over 500 years of colonization and oppression, resulting in the poisoning of our communities and land and the genocide of our peoples, do affirm and adopt these Principles of Environmental Justice:

1. Environmental justice affirms the sacredness of Mother Earth, ecological unity and the interdependence of all species, and the right to be free from ecological destruction.

2. Environmental justice demands that public policy be based on mutual respect and justice for all peoples, free from any form of discrimination or bias.

3. Environmental justice mandates the right to ethical, balanced and responsible uses of land and renewable resources in the interest of a sustainable planet for humans and other living things.

4. Environmental justice calls for universal protection from nuclear testing, extraction, production and disposal of toxic/hazardous wastes and poisons and nuclear testing that threaten the fundamental right to clean air, land, water, and food.

5. Environmental justice affirms the fundamental right to political, economic, cultural and environmental self-determination of all peoples.

6. Environmental justice demands the cessation of the production of all toxins, hazardous wastes, and radioactive materials, and that all past and current producers be held strictly accountable to the people for detoxification and the containment at the point of production.

7. Environmental justice demands the right to participate as equal partners at every level of decision-making including needs assessment, planning, implementation, enforcement and evaluation.

8. Environmental justice affirms the right of all workers to a safe and healthy work environment, without being forced to choose between an unsafe livelihood and unemployment. It also affirms the right of those who work at home to be free from environmental hazards.

9. Environmental justice protects the right of victims of environmental injustice to receive full compensation and reparations for damages as well as quality health care.

10. Environmental justice considers governmental acts of environmental injustice a violation of international law, the Universal Declaration on Human Rights, and the United Nations Convention on Genocide.

11. Environmental justice must recognize a special legal and natural relationship of Native Peoples to the U.S. government through treaties, agreements, compacts, and covenants affirming sovereignty and self-determination.

12. Environmental justice affirms the need for urban and rural ecological policies to clean up and rebuild our cities and rural areas in balance with nature, honoring the cultural integrity of all our communities, and providing fair access for all to the full range of resources.

13. Environmental justice calls for the strict enforcement of principles of informed consent, and a halt to the testing of experimental reproductive and medical procedures and vaccinations on people of color.

14. Environmental justice opposes the destructive operations of multinational corporations.

15. Environmental justice opposes military occupation, repression and exploitation of lands, peoples and cultures, and other life forms.

16. Environmental justice calls for the education of present and future generations which emphasizes social and environmental issues, based on our experience and an appreciation of our diverse cultural perspectives.

17. Environmental justice requires that we, as individuals, make personal and consumer choices to consume as little of Mother Earth's resources and to produce as little waste as possible; and make the conscious decision to challenge and reprioritize our lifestyles to insure the health of the natural world for present and future generations.

Adopted today, October 27, 1991, in Washington, D.C.
Source: First National People of Color Environmental Leadership Summit, Washington, D.C., October 24–27, 1991

Abbreviations and Acronyms

AEHR	Advocates for Environmental Human Rights
ASP	Africa Stockpiles Programme
BAN	Basel Action Network
BEA	British Environment Agency
BFR	Brominated flame retardant
CAI	Computer Aid International
CAP	Consumers' Association of Penang
CDC	Centers for Disease Control and Prevention (U.S.)
CEE	Central and Eastern Europe
CEJ	Coalition for Environmental Justice (in Central and Eastern Europe)
COHPEDA	Collective for the Protection of the Environment and an Alternative Development
CPA	Coalition Provincial Authority
CRCQL	Chester Residents Concerned for Quality Living
CSE	Center for Science and Environment
CTBC	Computer TakeBack Campaign
Danida	Danish aid agency
DDT	
DEET	
EIS	Environmental impact statement
EJNF	Environmental Justice Networking Forum of South Africa
EPR	Extended producer responsibility
ERRC	European Roma Rights Center
EU	European Union

EZD	DVDs manufactured for a single viewing and disposal
FAO	Food and Agriculture Organization
FASE	Foundation for Advancements in Science and Education
FDA	U.S. Food and Drug Administration
GAIA	Global Anti-Incinerator Alliance/Global Alliance for Incinerator Alternatives
GRRN	GrassRoots Recycling Network
HCWH	Health Care Without Harm
H-P	Hewlett-Packard
ICRT	International Campaign for Responsible Technology
IFI	International financial institution
ILO	International Labor Organization
IMF	International Monetary Fund
IPEN	International POPs Elimination Network
IPM	Integrated pest management
JOAP	Joint Oxfam Advocacy Program
LDC	Lesser developed country
LULU	Locally unwanted land use
MELA	Mothers of East Los Angeles
MMDA	Metropolitan Manila Development Authority
NEPSI	National Electronic Products Stewardship Initiative
NGO	Nongovernmental organization
NIMBY	Not in My Back Yard
OAS	Organization of American States
OECD	Organization for Economic Cooperation and Development
PAN	Pesticide Action Network
PANAP	Pesticide Action Network Asia and the Pacific
PANNA	Pesticide Action Network North America
PANUK	Pesticide Action Network United Kingdom
PBDEs	Polybrominated diphenyl ethers
PCBs	Polychlorinated biphenyls
PCR	People for Community Recovery
PET	Polyethylene terephthalate
PIRG	Public Interest Research Group
POPs	Persistent organic pollutants
POS	Political Opportunity Structure

PRCC	Plastics Recycling Corporation of California
RCRA	Resource Conservation and Recovery Act (U.S.)
REACH	Registration Evaluation Authorization of Chemicals (EU)
RoHS	EU Directive on Restrictions of Hazardous Substances
RTS	Return to sender
SAEPEJ	South African Exchange Program on Environmental Justice
SAP	Structural adjustment program
SMO	Social movement organization
SVTC	Silicon Valley Toxics Coalition
TCE	Texas Campaign for the Environment
TEAN	Taiwan Environmental Action Network
TNC	Transnational corporation
TSMO	Transnational social movement organization
UK	United Kingdom
UN	United Nations
UNFAO	United Nations Food and Agriculture Organization
UNEP	United Nations Environment Programme
UNICOR	Trade name for Federal Prison Industries (FPI)
USAID	U.S. Agency for International Development
USEPA	U.S. Environmental Protection Agency
WCED	World Commission on the Environment and Development
WEEE	EU Directive on Waste from Electrical and Electronic Equipment
WHO	World Health Organization
WMX	Waste Management Corporation
WSSD	World Summit on Sustainable Development
WTO	World Trade Organization
WWF	World Wildlife Fund

Notes

Chapter 1

1. This was an impressive accomplishment and a disturbing indicator of how much waste was being shipped into and burned in this community.

2. Gedicks 2001, Palmer 2005.

3. Heeten Kalan made this point in an interview with the author, 2005; see also Newell 2005.

4. Newell 2005.

5. Alston and Brown 1993; Pellow, Weinberg, and Schnaiberg 2001.

6. Adeola 2000; Critharis 1990; Clapp 2001; Frey 1995, 1998, 2001; Marbury 1995; Tiemann 1998.

7. Bullard 2000, 2005a,b; Bryant 1995; Camacho 1998; Cole and Foster 2001; Hurley 1995; Lerner 2005; United Church of Christ 1987.

8. Leading ecological modernization scholars include Arthur Mol, Gert Spaargaren, and David Sonnenfeld. See Mol 1995, 2003. The risk society thesis is primarily associated with the work of Ulrich Beck. See Beck 1992, 1995, and 1999. Leading treadmill-of-production scholars include Allan Schnaiberg and Kenneth Gould. See Gould, Schnaiberg, and Weinberg 1996; Schnaiberg 1980; Schnaiberg and Gould 2000; and Weinberg, Pellow, and Schnaiberg 2000. The work of Julian Agyeman and his colleagues suggests similar directions by exploring the need to link research on environmental inequality with theories of sustainability (see Agyeman, Bullard, and Evans 2003, and Agyeman 2005).

9. Others have made similar arguments from a standpoint that pays attention to the economic rather than racial dimensions of these inequalities. See, for example, Dunlap and Marshall 2006. For a response to this critique of the Eurocentrism of ecological modernization, see Mol 2003. The data I present in this study challenge Mol's theory and his defense on this point.

10. Gedicks 1993, 2001; McClintock 1995; Marable 1983; Melosi 2001; Schnaiberg 1980; Schnaiberg and Gould 2000; A. Smith 2005.

11. Increasingly we can observe this process occurring regionally among dominant nations and their close yet economically and politically less powerful neighbors. This dynamic roughly characterizes Japan's relationship with many Southeast Asian nations.

12. Tarr 1996, Melosi 2001.

13. Foster 2000, Schnaiberg 1980.

14. DuBois [1935] 1977, Goldberg 2002, Guinier and Torres 2002, Horsman 1981, Said 1979, West 1999.

15. A number of scholars have tackled the question of "color-blind society" versus "color-blind racism," including Bonilla-Silva 2003, Crenshaw 2000, and P. Williams 1997.

16. In the summer and fall of 2005, for example, several incidents occurred that underscore this point. One such moment was when the Mexican government commemorated an old popular cultural character, Memin Pinguin, on a national stamp. In a transnational racial incident, this state action caused an uproar among African American leaders like Jesse Jackson, as well as public intellectuals such as Carlos Muñoz, because Pinguin was a classic stereotypically African caricature, with exaggerated physical features (large lips, kinky hair), resulting in comparisons between Pinguin and U.S. images of Sambos and other controlling racist symbols. Mexican president Vicente Fox, along with many other Mexican leaders, denied both that the Pinguin character and stamp were racist and argued that Mexico was a nation free of racism (see Althaus and Hegstrom 2005). Of course, that claim stems from the history of Mexican racial formation officially viewed as *mestizaje,* or a mixture of racial, ethnic, and cultural groups, that create one people unburdened by racism. This claim is belied by the daily confrontations and insults of racism that Mexicans of both indigenous and African descent experience, which in the former case contributed to the Zapatista uprising in 1994. Another major moment occurred when, in France, after two African youth died during a police chase, thousands of African and Muslim youth rioted across that nation, protesting racism at the hands of the state and police forces, and in schools and economic institutions (see M. Moore 2005). The charge was that these youth of color had no future and were perpetually treated like foreigners, even when they were born in France. President Jacques Chirac, Prime Minister Dominique de Villepin, and many white French citizens vehemently countered with declarations that France was a color-blind nation that made no distinctions between racial or national groups. These claims are belied by the massive unemployment, residential segregation, and near total exclusion from the French parliament that African and Arab persons in France experience. Finally, in late fall 2005, riots broke out in Australia, when tensions emerged between youths of Arab descent and Euro-Australians in beach towns. For days, white male youths roamed Australian cities, beating people who "looked Arab" (see Associated Press 2005a). Prime Minister John Howard's response was predictable: he declared repeatedly that Australia was a color-blind society. Taken together, these and many other moments make it plain that the discourse of color-blindness is a global rhetorical device that heads of state and insti-

tutions are deploying to mask the reality of deeply structured and violent racisms around the world.

17. Keck and Sikkink 1998, p. 18. *Venue shopping* is a term borrowed from law, which involves presenting a case and searching for a receptive political venue. "The recent coupling of indigenous rights and environmental issues is a good example of a strategic venue shift by indigenous activists, who found the environmental arena more receptive to their claims than human rights venues had been" (ibid.). But scholars and activists should never assume that just because a policy decision is moved from a national to an international arena, its resolution will occur more easily. In fact, Mark Ritchie argues that the U.S. and other governments are only too happy to move certain issues to the international arena of multilateral trade institutions (like the North American Free Trade Agreement) precisely because these institutions are less open to societal influence (Khagram, Riker, and Sikkink 2002, 9.

18. Keck and Sikkink 1998; McAdam, Tarrow, and Tilly 2001; Smith, Chatfield, and Pagnucco 1997; Tarrow 1998.

19. Almaguer and Moore et al. remind us of the ever-changing faces and contours of racism through time (Almaguer 1994; Moore, Kosek, and Pandian 2003). M. Goldman (1998) demonstrates how imaginaries and practices have shaped the natural landscapes around the globe in continuous struggles among corporations, states, workers, and activists fighting over the ownership, size, and the very notion of the commons.

20. Marbury 1995.

21. Porterfield and Weir 1987.

22. Marbury 1995.

23. Schnaiberg and Gould 2000, Yearley 1996.

24. Summers 1991.

25. Hilz notes that economics offers a cold calculus that supports such practices: "From a microeconomic perspective of an exporting country, the practice of waste export is efficient. Large sums of money may be saved, by avoiding waste treatment costs, such as incineration, landfilling, or recycling, and liability costs resulting from possible future damages due to inadequate waste management, and insurance coverage for hazardous waste facilities. Distortions resulting from different regulatory standards, externalities, and the transfer of risk support the argument that hazardous wastes should not be handled as economic goods" (Hilz 1992, 5).

26. Goldsmith and Mander 2001, United for a Fair Economy 2002.

27. DuBois 1935, Newell 2005, Winant 2001.

28. Critharis 1990, Marbury 1995, Tiemann 1998, Yang 2002.

29. Adeola 2000; Frey 1994, 1998, 2001.

30. Clapp 2001.

31. Greenpeace International is the organization involved in transnational waste trade environmental justice conflicts; Greenpeace USA is the arm of the organization working with U.S. domestic environmental and environmental justice groups.

32. Hilz 1992.

33. Ibid., O'Neill 2000.

34. Adeola 2000, Moyers 1990, Mpanya 1992, Porterfield and Weir 1987.

35. Gbadegesinn 2001, 201.

36. Mpanya 1992, 212.

37. Kovel 1984.

38. Hilz 1992, 5.

39. Marbury 1995.

40. Porterfield and Weir 1987, 12.

41. Bullard 2000.

42. Lipsitz 2001, Pulido 2000.

43. Frey 2001.

44. This has also occurred in the case of mining and other resource extractive industries: "In 1992, the number of U.S. and Canadian mining corporations exploring or operating in Latin America doubled from the year before" (Charlier 1993). Latin America quickly became the leading nation for new mining investment, followed by Africa and Asia. What accounts for the dramatic increase in mining investment? Mining executives complained about stiffer environmental regulations and the long delays in the mine permitting process because of objections from environmentalists in the United States and Canada (Gedicks 2001).

45. Gedicks 2001, 197. In this case, Gedicks is speaking primarily about natural resource extraction, but the same principle applies to environmental harm through waste dumping.

46. Szasz 1994.

47. Hilz, 1992, 43. Klaus Toepfer, the minister for the environment, nature protection, and nuclear safety in what was at the time called West Germany referred to the waste sector as being in a "state of emergency" (Hilz 1992, 42).

48. Pellow 2001.

49. For good overviews of this literature, see Brulle and Pellow 2006 and Szasz and Meuser 1997.

50. Bullard, 2000, Cole and Foster 2001, Gottlieb 1993, Pulido 1996, Roberts and Toffolon-Weiss 2000.

51. Alston 1990.

52. Agyeman 2005; Bullard 2000, 2005a; Bullard, Johnson, and Torres 2000; Bullard and Wright 1993; S. Fox 1992; Gedicks 1993; Hurley 1995; LaDuke 1999; Pellow 2002; Pellow and Park 2002.

53. Bullard 2000; Lavelle and Coyle 1992; Morello-Frosch, Pastor, Porras, and Sadd 2002.

54. Been 1993a, 1993b.

55. Faber and Krieg 2001, Massey and Denton 1993.

56. Cole and Foster 2001.

57. Bryant 1995, Gottlieb 1993.

58. Bullard 2000; Bullard, Johnson, and Torres 2000.

59. Gedicks 1993, Gedicks 2001, Deloria 1985, LaDuke 1999, Smith 2005.

60. Gottlieb 2001, Levenstein and Wooding 1998.

61. For an early articulation of the concept of institutional racism, see Turé and Hamilton [1968] 1992.

62. Bullard 2000.

63. Seager 1994.

64. Brown and Ferguson 1995, Pellow and Park 2002.

65. I am indebted to Kenneth Gould for this observation, which he made at the Sociologists Without Borders conference in Philadelphia, August 2005.

66. Adeola 2000; Agyeman, Bullard, and Evans 2003; Bullard 2005; McDonald 2002; Taylor 1995; Westra and Lawson 2001.

67. Alexander 1990, 16 (see Seippel 2002, 198).

68. Mol 1995, 1996; Mol and Sonnenfeld 2000; Spaargaren 1997; Spaargaren and Mol 1992.

69. Fisher and Freudenburg 2001, Sonnenfeld 2000, Spaargaren 1997.

70. Mol 1995, 37.

71. Seippel 2002. This argument is reminiscent of (and challenges) Audre Lorde's caution that one cannot expect to dismantle the master's house with the master's tools (Lorde 1984).

72. Mol 1995, Spaargaren 1997, Weale 1992, Huber 1982.

73. Yearley 2002, 282.

74. Garcia-Johnson 2002.

75. Gould, Schnaiberg, and Weinberg 1996; Hooks and Smith 2004; Schnaiberg 1980; Schnaiberg and Gould 2000; Weinberg, Pellow, and Schnaiberg 2000.

76. O'Connor 1991.

77. Schnaiberg and Gould 2000.

78. Gould, Weinberg, and Schnaiberg 1996.

79. Beck 1992, 1999.

80. Bullard 2000.

81. Hurley 1995.

82. Seippel 2002.

83. Benton 2002, 261.

84. For an outstanding study of the tension between citizens and experts in defining and solving environmental problems see F. Fischer 2000.

85. See http://www.globalresponse.org. Global Response, an international environmental education and action network, uses Phillips's quote as one of its major slogans.

86. This point Beck makes has parallels with the political economic opportunity structure model I present in this chapter and chapter 2.

87. Beck 1992, 35.

88. Beck 1995; see also Beck 1999.

89. Beck 1992.

90. I am indebted to Robert Brulle for his insights and guidance on Beck's writings.

91. Melosi 2001, 67.

92. See Hilz 1992.

93. Hilz 1992, 10.

94. Ashford 1992, in Hilz 1992, xiii.

95. Krishnamoorthy 2003.

96. Raffensperger and Tickner 1999.

97. G. Cohen 1997.

98. Goldman 1991, 196–197.

99. Environment News Service 2004c.

100. Centers for Disease Control and Prevention, 2003.

101. Lunder and Sharp n.d.

102. Environmental News Service 2003e.

103. Porterfield and Weir 1987, Weir and Shapiro 1981.

104. Beck 1991; Pellow and Brulle 2005. It is true that one of the reasons for the increased reported levels of hazardous waste produced in the United States and other nations (such as Germany) has been a reclassification of certain materials now known to be more hazardous than previously thought. But it is also the case that there has been a media shift from dumping (primarily) at sea and in the air to the land. When ocean and air dumping were more heavily regulated, it placed greater pressure on states to find other media (Hilz 1992, 31, 33–34). This media shift is reflected in the following seemingly contradictory headline: "Overall Toxic Releases Down, Hazwaste Up in 2000." The article reports that while "overall toxic releases [from U.S. industries] are down, quantities of hazardous wastes are up. The report documents a 25 percent jump in the amount of hazardous wastes disposed of through landfills, incinerators and other methods in the year 2000" (Environmental News Service 2002c). This "contradiction" is a result of shifting pollution from air to the land, something that we have ob-

served in U.S. environmental practices of decades past (Pellow 2000). In fact, another news story reported the events more accurately, with the following headline: "North America Shifts Pollution from Air to Land" (Environmental News Service 2002b).

105. Liptak 2004.

106. Lee 2004.

107. Ibid., emphasis added.

108. United Nations 2005.

109. See the RoHS [Restriction of Hazardous Substances] and WEEE Directives [Waste from Electrical and Electronic Equipment] in the EU; Royte 2006.

110. Author's interview with Von Hernandez, February 24, 2004.

111. Pegg 2004.

112. Environmental News Service, 2004a; see Redefining Progress 2004.

113. Environmental News Service 2002a.

114. Ibid.

115. Hilz 1992.

116. OECD 2001, 137, 314 (see Montague 2003, 19–41).

117. Thanks to Eric Klinenberg for making this astute observation.

118. World Commission on Environment and Development 1987, 35.

119. United Nations 1973; Tolba 1988, 7.

120. Other scholars have made similar points from a world systems and class analytical perspective. See, for example, Goldfrank, Goodman, and Szasz 1999.

121. University libraries I used for this research include the main library and law libraries at the University of California, Berkeley, Geisel Library and the International Relations/Pacific Studies Library at the University of California, San Diego, and the main library at the University of Colorado at Boulder. Transnational environmental justice and human rights organizations and networks such as Basel Action Network, European Roma Rights Centre, Global Alliance for Incinerator Alternatives, Global Response, Health Care Without Harm, International Campaign for Responsible Technology, Pesticide Action Network Asia Pacific, Silicon Valley Toxics Coalition, and the Transatlantic Initiative on Environmental Justice provided critical documents, reports, and other data.

122. These include the Global Symposium for a Sustainable High-Tech Industry (San Jose, California, fall 2002); Encuentro Border Environment Conference (Tijuana, B.C., Mexico, spring 2003); "Just in Time: Bringing Ethical Standards to the Electronics Sector" roundtable, sponsored by the Catholic Agency for Overseas Development (spring 2004); the Environmental Justice Summit II (Washington, D.C., fall 2003); "Labor Rights and Environmental Justice in the Global Electronics Industry," Workshops (held in San Diego, the Netherlands, and Thailand, summer-fall 2003); and the Transatlantic Initiative on Environmental Justice (Budapest, Hungary, fall 2005).

123. The 2004 American Sociological Association annual meetings in San Francisco featured the theme "Public Sociologies," which brought the questions of the relevance, impact, purpose, and meaning of scholarly research to the foreground of the conference. The 2007 ASA meeting theme is "Is Another World Possible?" echoing the slogan of the World Social Forum.

124. Feagin and Vera 2001.

125. Marx 1962.

126. DuBois 1977.

127. Feagin and Vera 2001, 264.

128. Philo and Miller 2001, 79.

129. These include the Silicon Valley Toxics Coalition, the International Campaign for Responsible Technology, Global Response International Environmental Education Network, People for Community Recovery, the Santa Clara Center for Occupational Safety and Health, worksafe!, the Transatlantic Initiative on Environmental Justice, and the San Diego Foundation.

Chapter 2

1. Winant 2001.

2. Ibid.

3. Sale 1990; Dussel 1998.

4. For example, Merchant 1980, Kovel 1984, Mills 1997, Okihiro 1994, Sale 1990, Silva 2007, Todorov 1984, Toynbee 1934.

5. Kovel 1984, liv. This racialized process also has strong parallels with the development of patriarchy in modernity. For Carolyn Merchant and many other ecofeminist scholars, nature is also largely viewed as a degraded, undeveloped, feminine entity in need of masculine guidance and dominance. Thus, gender domination is a core part of modernity and environmental manipulation.

6. DuBois 1977, 15.

7. Kovel 1984, N. Smith 1996. "Conquest" in late twentieth- and early twenty-first-century urban America might be characterized as what is commonly referred to as "gentrification."

8. Santa Ana 2002.

9. Appiah 2006, Appadurai 1996, Gilroy 2001. I am indebted to John Marquez for making this point.

10. Pellow and Park 2002; Smith, Sonnenfeld, and Pellow 2006.

11. Gedicks 1993, 2001; LaDuke 1999; Agbola and Alabi 2003; Wright 2005.

12. For example, the U.S.A. Patriot Act, Operation Gate Keeper at the U.S./Mexico Border, and Britain's "Managed Immigration" policies (see Bacon 2004). Scholars I would place in this critical school of globalization and race studies

include Andrew Barlow, Howard Winant, Ward Churchill, Al Gedicks, Cornel West, Manning Marable, and Peter Newell (see Marable 2004 and Newell 2005).

13. Hannaford 1996, Magubane 1990, Rodney 1981, R. Williams 1990.

14. See, for example: Mander and Goldsmith 2001, Schaeffer 2003, Stiglitz 2003, Smith and Johnston 2002.

15. Pellow 2001.

16. Barlow 2003, 64.

17. United Nations Development Program 1999, 3.

18. Wallach and Woodall 2004, 9–10.

19. Roberts and Grimes 2002, 177.

20. Winant 2001, 305. This racialization of world inequality has led to the development of what Charles W. Mills calls global white supremacy, that is, a series of racial orders structured along the axis of white (or European) and nonwhite (or non-European) in the world system (Mills 1997). Amy Chua's book *World on Fire* (2003) adds further depth to the study of global racism, as she examines the invidious political and economic gulfs between racial/ethnic populations within nations around the world. Specifically, there are often ethnic groups that are the statistical minority in many countries and yet control the lion's share of the nation's economic wealth. Chua argues forcefully that, contrary to the modernists' claims, it is precisely the free market globalization policies of the North that produce and exacerbate these racial and ethnic tensions around the world, revealing continued racism and often breeding violent backlashes when members of disempowered ethnic majorities rebel against privileged minority groups.

21. Bonilla-Silva 2001, Barlow 2003.

22. Winant 2001.

23. Schnaiberg 1980, Schnaiberg and Gould 2000.

24. See Barnet and Muller 1974, 123–147; Starr 2000. Patricia Hill Collins argues that the rise of privatization in public programs within the United States is rooted in a racist drive to reduce public funding for otherwise public institutions and functions that are often inaccurately associated with programs designed to benefit people of color. I would argue that we see this logic on a global scale as well, in the move from colonialism to neocolonialism (Collins 2005).

25. Some examples of anticorporate, antiglobalization literature that sidesteps racism's role in shaping and maintaining economic and cultural globalization are Mander and Goldsmith 2001, Nace 2005, Wallach and Woodall 2004, Manheim 2001, Ritz 2001. Sources on antiglobalization—or global justice—movements that more consciously locate race and racism in this history include Starr 2000 and Prokosch and Raymond 2002.

26. Goldberg 2002, 102–103, discusses the debates around "relative autonomy" of states and capital, but does not develop the point that capital coproduces racial hierarchies with the state. In a brilliant study of racialized state violence, Joy James

(1996) also pays little attention to the significant role that private power plays in guiding, supporting, and sustaining such systems of domination.

27. Boggs 2000, 11, 68. In a lesser-known text, Andrew Hacker (1965), who is much better known as a race theorist, wrote more than forty years ago about the rise of "corporate form" that had emerged "as the characteristic institution of American society."

28. West 1997, 8–10.

29. Guinier and Torres 2002, 11, emphasis added.

30. Ibid., 12, emphasis added.

31. Douglass 1881. Cf. in Feagin, Vera, and Batur 2001, 16–17, emphasis added.

32. Moore, Kosek, and Pandian 2003, 9.

33. F. Cohen 1953, 348–390; Mosley 1992, p. 87.

34. The logo for a conference on environmental justice at Macalester College in St. Paul, Minnesota, read: "Demand Environmental Justice! Racism Is Toxic" (Seventh Annual American Studies Conference, Macalester College, February 24–26, 2006).

35. Bullard 1992.

36. Foster 2000.

37. Wilson 1990.

38. Sennett and Cobb 1993.

39. Faber 1998.

40. Omi and Winant 1994, Goldberg 2002.

41. Marable 1983.

42. The general exception is that they must operate within the parameters of the law, although corporations generally have a much stronger hand in shaping those laws than the citizenry.

43. Barlow 2003, 12.

44. Texaco 1994. This was the case of *Roberts* v. *Texaco,* brought on behalf of fourteen hundred company employees.

45. Fenton Communications 1999.

46. Feagin, Vera, and Batur 200. Note that since Europeans do not have a monopoly on racism, here we are discussing white racism, which is a particularly pernicious and global form of inequality that has affected more people on the planet than any other form of racism. There are, of course, other forms of racism that occur between and among people outside of European and American contexts.

47. Pellow and Park 2002.

48. Westra 1998.

49. Beck 1992, 35.

50. Newell 2005, 80–81; Schnaiberg and Gould 1994.

51. Foster 2000, 74.

52. Hall 1978

53. Takaki 1998, Vo 2004.

54. According to the National Asian Pacific American Legal Consortium (NAPALC), there was a spike in anti-Asian hate crimes and violence in 2001, following the attacks of September 11, 2001, with South Asians, particularly Indians and Pakistanis, receiving a disproportionate level of attacks, due to the fact that many South Asian men are Sikhs and wear turbans and long beards, two of the symbols that have come to be associated with terrorism in the media. That year there were 597 reported bias-motivated hate crimes against Asian Pacific Americans, with people using threats, vandalism, arson, rape, and assaults with baseball bats, metal poles, and guns. Unfortunately, the "9/11 thesis" cannot account for the overall trend because this was the second spike in four years, as there was a 13 percent rise in anti-Asian hate crimes between 1998 and 1999 (NAPALC 2001). These crimes are part of a general anti-Asian bias in U.S. history punctuated by riots, massacres, lynching, and legislation designed to control, oppress, and expel members of these groups since the mid-nineteenth century (Lowe 1996).

55. Espiritu 2003, 47.

56. Guinier and Torres 2002, 30.

57. McCarthy and Zald 1977.

58. Guidry, Kennedy, and Zald 2000, 3.

59. Bhabha 2004, Spivak 1999, Vinayak 2000.

60. Guidry, Kennedy, and Zald 2000, 18.

61. McAdam 1999.

62. Tarrow 1998, 2.

63. Ibid.

64. Smith, Chatfield, and Pagnucco 1997, 42; Keck and Sikkink, 18; McCarthy 1997, 257.

65. Tarrow 1998, 189

66. For example, renowned global justice scholar-activists Martin Khor from the Third World Network (Penang, Malaysia), Vandana Shiva (Research Foundation for Science, Technology, and Natural Resource Policy, New Delhi, India), and Walden Bellow (Focus on the Global South, Philippines).

67. Heeten Kalan, a South African man, works for groundWork USA (a U.S.-South African NGO) and lives in the Boston area; Urungu Houghton and Njoki Njoroge Njehu, both Kenyans, work at Action Aid USA/Oxfam and 50 Years Is Enough, respectively. Njehu has worked with women's groups and the Greenbelt Movement in Kenya for over a decade. She grew up learning from the work of Kenyan women, especially her mother, Lilian Njehû, a grassroots and community activist. Before joining the 50 Years Is Enough Network, she worked at Greenpeace International for three years, focusing on the international toxic trade and on biodiversity and

oceans issues. Houghton has lived and worked in the North for years and lectures regularly at international fora and universities in the United States and Europe.

68. Between 1953 and 1993, the number of TSMOs rose from fewer than two hundred to more than six hundred (Smith, Chatfield, and Pagnucco 1997).

69. Schnaiberg and Gould 1994.

70. Winant 2001.

71. See, for example, Boli and Thomas 1999; Florini 2000; Risse, Ropp, and Sikkink 1999; Smith, Chatfield, and Pagnucco 1997; Keck and Sikkink 1998; Stiles 1998. International norms can be defined as shared expectations or standards of appropriate behavior that are accepted by states and intergovernmental organizations that can be applied to states, intergovernmental organizations, and nonstate actors (Khagram, Riker, and Sikkink 2002, 14).

72. For example, there are numerous national or global "day of action" events coordinated across many cities and nations by antiwar, feminist, and environmental and environmental justice TSMOs such as International A.N.S.W.E.R. (Act Now to Stop War and End Racism), the Rainforest Action Network, the Global Anti-Incinerator Alliance (GAIA), and the International Rivers Network.

73. Keck and Sikkink 1998; Smith, Pagnucco, and Chatfield 1997.

74. Khagram, Riker, and Sikkink 2002, 16.

75. Gould, Weinberg, and Schnaiberg 1996.

76. Yearley 1996, 87; see also Palmer 2005.

77. Florini 2000, 6.

78. Roberts and Toffolon-Weiss 2001.

79. Fox and Brown 1998.

80. Florini 2000.

81. Infamous examples include the experience of the Huaorani indigenous peoples with the U.S.-based Natural Resources Defense Council, in which the latter negotiated a deal with the government of Ecuador and oil companies to protect certain lands and sacrifice native lands, all without full consultation or permission from Huaorani leaders (see Gedicks 1993).

82. Keck and Sikkink 1998, 206–207.

83. People against Foreign NGO Neocolonialism 2003.

84. See Proceedings from the First National People of Color Environmental Leadership Summit, Washington, D.C., October 1991.

85. Eisinger 1973.

86. Kitschelt 1986, McAdam 1999, Tarrow 1998.

87. van der Heijden 2006.

88. McCarthy 1997, 255.

89. Keck and Sikkink 1998, 7.

90. For examples of this perspective, see McAdam 1996, 34; Marks and McAdam 1996.

91. Boudreau 1996, 176.

92. Mander and Goldsmith 2001, Wallach, and Woodall 2004.

93. United Nations Conference on Trade and Development 1998.

94. Massive riots against the IMF and the World Bank in global South nations over the past thirty years are a strong indicator of this. Recently citizens of Argentina blamed their economic woes on these twin institutions, and many antiglobalization protests around the world since 1999 have targeted the IMF, World Bank, and World Trade Organization as primary facilitators of corporate-led globalization.

95. Pellow 2001.

96. While the power of private capital is more intense today than it has been in decades, the problem of corporate dominance is not new. In 1863, during the middle of the U.S. Civil War, Abraham Lincoln observed, "I have the Confederacy before me and the bankers behind me, and I fear the bankers most." Decades later, President Franklin D. Roosevelt similarly warned that "the liberty of a democracy is not safe if the people tolerate the growth of private power to a point where it becomes stronger than their democratic state itself. That, in its essence, is fascism—ownership of government by an individual, by a group, or by any controlling power." In the 1930s, right-wing political leaders aligned themselves with industrialists, creating fascist states that destroyed both human rights and ecological quality. Benito Mussolini himself once argued that "fascism should really be called 'corporatism'" (Kennedy 2003). Dwight D. Eisenhower made what is perhaps the most famous cautionary note on this topic in 1961 when he warned the country of the ominous and rising power of the "military-industrial complex."

97. I am indebted to Ted Smith for this point. For example, the Computer Take-Back Campaign exploited the divisions within the electronics industry over how to finance product recycling systems, to push a group of the largest firms toward more progressive policies (see chapter 6).

98. One scholar who proposes a model closest to the political economic opportunity structure concept is Sigrun I. Skogly, a legal scholar at Lancaster University in the United Kingdom, who urges the adoption of a broader definition of international political opportunity structure that includes "international private actors" such as "international banking and transnational corporations" (Skogly 1996, Alger 1997, 272.). This is a rare theoretical acknowledgment of the power of TNCs in producing a different kind of political process—a political economic process—than social movement scholars have recognized. Skogly's proposal seems not to have taken root, however, and this is the task before us in this book.

99. Smith, Pagnucco, and Chatfield 1997, 70.

100. Konefal and Mascarenhas 2005, Schurman 2004.

101. Keck and Sikkink 1998, 200.

102. See Brenner 2002, Ong 2004.

103. Bakan 2004.

104. See Collins 2005. This was the core of the battle over California's Proposition 187, which would have restricted public services like education and heath care by excluding undocumented immigrants' access. In Scandinavian nations, often celebrated as models of democratic socialism, we are now witnessing a similar restriction of public services as new immigrant populations increase and ethnocentrism rises.

105. Bakan 2004, O'Rourke 2005.

106. Gould 2005.

107. Hacker 1965, Isin 1998, McDermott 1991.

108. A favorite example that the left uses today is the George W. Bush administration, which has deep and enduring connections to several environmentally destructive industry sectors, especially the energy and petroleum industry.

109. This may reflect what organizational behavior theorists have called institutional isomorphism, where organizations adopt behaviors of other, successful organizations (DiMaggio and Powell 1983).

110. Robert Alvarez refers to these workers as "'offshore nationals' who say their job is to protect American farmers." (Alvarez 2006).

111. Isin 1998.

112. Greenpeace USA 2006b, emphasis added. The popular "Incorporated States of America" flag is a U.S. flag that features the logos of several TNCs in the upper left corner where the stars representing fifty states would normally appear. A related piece of progressive-left messaging is a sticker that reads, "Since when did unquestioning obedience to corporate interests become patriotic?" (Duck and Cover 2003).

113. Khagram, Riker, and Sikkink 2002, 15.

114. Kriesberg 1997, 5.

115. McCarthy 1997, 256.

116. Sachs 1995, 45.

117. Gedicks 2001, 197–198; Keck and Sikkink 1998, 27.

118. Tarrow 1998 argues that transnational movements must engage in sustained disruptive actions against targets in order to be successful.

119. Hochshild 2005, Keck and Sikkink 1998.

120. Barlow 2003, 147–148.

121. Amin 1989, Rodney 1981. Of course, regional colonization also occurred around the world among non-Europeans in Latin America, Africa, and Asia, often supported by racism. Japan and China are among some of the most powerful examples. Over the centuries, both states have occupied and colonized many nations in the Asia Pacific region (I. Chang 1998, Human Rights Watch 1999, Johnson 2003).

122. Winant 2001.

123. "Official statements of global opposition to racism, colonialism, and xenophobia appeared early in the development of the modern framework of international cooperation, such as in the 1948 Universal Declaration of Human Rights. . . . The UN Conference on Racism, Xenophobia, and Related Problems, held, appropriately enough, in Durban, South Africa, in 2001, marked an important stage in the development of human rights by providing an international setting to develop a global antiracist agenda for the first time" (Barlow 2003, 150). For an outstanding study of how international human rights politics shaped the civil rights movement in the United States, see Anderson 2003.

124. Keck and Sikkink 1998, 215.

125. The World Bank's emphasis on fighting corruption through the promotion of "good governance" in the South—and environmental NGOs' applauding of this approach—is a good example (Wolfowitz 1996).

126. The degree of bribes and money laundering that northern institutions and individuals offer for southern institutions is extraordinary (see Hawley 2005 and Kauffman 2004). With regard to undemocratic practices in the North, consider the following example. The majority of U.S. citizens polled were opposed to the North American Free Trade Agreement, yet Congress ratified it anyway. The majority of Britons, Australians, and Spaniards were opposed to the U.S.-led war in Iraq (begun in 2003), yet their governments went forward with the invasion anyway. The majority of U.S. citizens would prefer stable jobs and unpolluted neighborhoods, but the government and corporations continue to make this an elusive dream by favoring the global sourcing of manufacturing jobs and gutting environmental protection regulations (domestically and through free trade agreements). The United States is the only industrialized nation that has not abolished the death penalty despite repeated calls by its own citizens and global human rights groups to do so. The Bush White House has refused to honor or ratify numerous international agreements respected by much of the world community (e.g., the Kyoto Protocol, the Basel Convention). That same administration has repeatedly withheld critical information from the public, the press, independent commissions, and the courts over matters of great importance regarding the erasure of civil liberties and the overhaul of many basic regulations. In a single week in April 2004, the White House argued two cases before the U.S. Supreme Court. The first was an effort to keep secret the records of Vice President Dick Cheney's energy task force meetings with private energy corporations, which are believed to have significantly shaped subsequent national policies. The second was an effort to maintain the practice of detaining U.S. citizens believed to be involved in terrorist activities indefinitely, with no criminal charges against them, and with no legal representation. In 2004, African American leaders from Washington, D.C., went before the Organization of American States (OAS) protesting the fact that the population of that city has no congressional representation in the capital and therefore is a violation of international law. The OAS declared that this claim was valid and ruled in their favor—a measure that was promptly ignored by the U.S. Congress. From

the Palmer raids of the early twentieth century to the Federal Bureau of Investigation's Counter Intelligence Program, secret wars against countless domestic dissident groups, to the Central Intelligence Agency's support of assassinations of heads of state and of coups d'état around the world (even in democratically elected nations), the United States has earned the unenviable reputation among many scholars, activists, and citizens as the leading Rogue State (Blum 2000). See also Blum 1995; N. Chang 2002; Chomsky 1993; Churchill and Vander Wall 2001; Johnson 2000, 2004.

127. Guidry, Kennedy, and Zald 2000, 5.

128. After the highly controversial and irregular presidential election of 2000 in the United States, Cuban president Fidel Castro offered to serve as an observer of the next election to ensure that the process was "free and fair" according to international standards. Washington was not amused by this overture. Despite overwhelming evidence of blatant fraud and deception in that election, no one has been brought to justice.

129. See Boggs 2000, Mohanty 2003.

Chapter 3

1. As there are numerous transnational environmental networks today, this exploration is necessarily limited. For example, because the Indigenous Environmental Network (IEN) has received extensive consideration by other scholars, I exclude it from this discussion (see Cole and Foster 2001, LaDuke 1999). The transnational networks I focus on are also highly integrated and collaborate with each other, while IEN tends to be less central to these groups' efforts. The same could be said of a number of other networks.

2. I present these networks in alphabetical order.

3. Basel Action Network Web site: http://www.ban.org.

4. Leonard 2005a, b.

5. Author interview with Chet Tchozewski, March 11, 2002. Freedom Summer is the name of the massive voter registration effort in 1964 led by major civil rights organizations and conducted mainly by brave young African American and white students.

6. http://www.greenpeace.org/international/about.

7. Health Care Without Harm Web site: http://www.noharm.org.

8. From the ICRT mission statement at http://www.svtc.org/icrt/index.html.

9. Smith, Sonnenfeld, and Pellow 2006, 5.

10. GAIA's global co-coordinators at the time, Von Hernandez and Ann Leonard, were present and actively involved at ICRT's founding meeting.

11. See Pesticide Action Network Web site: http://www.panna.org.

12. Author interview with Palmer. See also Gould 2003.

13. The Union Carbide plant in Bhopal, India, experienced a massive chemical accident in December 1984, killing thousands of people. This is believed to be the worst chemical disaster in history. Dow acquired Union Carbide in 2001, inheriting this liability (see Doyle 2004).

14. Author interview with Sathinath Sarangi, February 2003.

15. Author interview with anonymous global South activist, 2005. I maintain this person's anonymity because of the highly sensitive nature of this quotation.

16. Author interview with Bobby Peek, August 24, 2004.

17. Author interview with Heeten Kalan, October 5, 2005.

18. This debate has been ongoing for some years among scholars (see Anderton, Anderson, Oakes, and Fraser 1994; Bryant and Mohai 1992; and Krieg 1998) and is one that I and other scholars have noted is unproductive since environmental inequalities are generally the result of both race and class exploitation (Faber and Krieg 2001, Pellow and Brulle 2005). In this book, however, I focus primarily on racialization because it is a theoretical space that has yet to be fully explored in the literature.

19. Author interview with Romeo Quijano, November 2004.

20. Author interview with Jim Puckett, March 5, 2002.

21. Author interview with Heeten Kalan, December 23, 2003.

22. Author interview with Ann Leonard, March 5, 2002.

23. Author interview with Nityanand Jayaramand, April 9 and 10, 2002.

24. Author interview with Peek.

25. Author interview with Puckett.

26. Anderson 2003.

27. Gordon and Harley 2005, Cole 2002.

28. Harden 2005.

29. Bullard 2005a, b.

30. Author interview with Quijano.

31. Author interview with Von Hernandez, February 24, 2004.

32. Author interview with Puckett.

33. Author interview with Francis dela Cruz, February 12, 2004.

34. Author interview with Puckett.

35. Author interview with Leonard. The Multinationals Resource Center is affiliated with the *Multinational Monitor Magazine,* a publication that focuses on the socially, economically, and environmentally harmful practices of transnational corporations.

36. Author interview with anonymous activist from South Asia, January 2004. I maintain this person's anonymity because of the highly sensitive nature of this quotation.

37. Gedicks 1993. In 1990, several U.S.-based environmental justice movement groups sent open letters to the leadership of the "big ten" environmental organizations protesting their insensitivity to the struggles of poor people and people of color, as evidenced in their lack of ethnic diversity in their membership, staff, and boards of directors and as embodied in their campaigns that often sacrificed the economic needs of vulnerable peoples in favor of narrow ecological goals.

38. Author interview with anonymous activist from South Asia, January 2004. I maintain this person's anonymity because of the highly sensitive nature of this quotation.

39. Author interview with Paula Palmer, August 2003. This quote also appeared in Palmer 2005.

40. Author interview with Annie Leonard, March 5, 2002.

41. Author interview with Palmer.

42. For writings on white privilege and the problem of whiteness, see Lipsitz 2005.

43. Author interview with Palmer.

44. Author interview with Kenny Bruno, March 7, 2002.

45. Author interview with Peek.

46. Author interview with Ravi Agarwal, April 1, 2002.

47. Author interview with Peek.

48. Author interview with Kalan, December 23, 2003.

49. Author interview with Kalan, December 23, 2003.

50. Keck and Sikkink 1998.

51. Bobby Peek has received international acclaim for his campaigning work in the South Durban industrial basin concerning toxic waste and polluting industries. He is a recipient of the prestigious Goldman Environmental Prize and, through groundWork, organizes communities internationally throughout Africa. Many groundWork staff members and trustees served on South Africa's Truth and Reconciliation Commission, so there are strong links to apartheid era activism. The organization works on air quality, medical waste and incineration, and industrial landfill waste issues, among others. GroundWork is active in international environmental justice networking and is affiliated with Health Care Without Harm, IPEN, BAN, and Oilwatch International.

52. Author interview with Kalan, December 23, 2003.

53. Since 1969, the CAP has advocated on behalf of consumers and residents concerning environmental protection, food safety, and harmful development projects. Speaking to the political economic opportunity structure framework, Mageswari Sangaralingam declared that CAP is "a fearless advocate of the people's right to a healthy and sustainable environment. It has championed the interest of citizens and communities against corporate lies and insatiability, toxic pollution and unsustainable use and management of land and other natural resources" (author interview with Mageswari Sangaralingam, February 26, 2004).

54. Author interview with Sangaralingam.

55. Ibid.

56. Ibid.

57. Author interview with Nityanand Jayaraman, March 5, 2002.

58. Snow and Benford 1992.

59. Author interview with Jayaraman.

Chapter 4

1. Mills 2001.

2. See Crowe 1996, Kovel 1984, Mills 1997, Santa Ana 2002.

3. Higgins 1994 (cf. Sze 2007).

4. For a celebrated journalist's perspective supporting this ideology, see Friedman 2005. For a lauded academic's version see Wilson 1990. And for a politician's model of social Darwinian policy, one need look no further than George W. Bush's tax cuts for the very wealthy in the Economic Growth and Tax Relief Reconciliation Act of 2001.

5. See Summers 1991, for example. Also, as Kenny Bruno of Greenpeace stated with regard to Philadelphia's incinerator ash being dumped in Haiti: "If everybody had to dispose of all their waste within a mile of their home, the incentive to reduce it would be far, far greater. If you could send it to the moon, you would have no reason to reduce it. And unfortunately, to some people, Haiti is like the moon" (Schwartz 2000). This reveals an out-of-sight, out-of-mind mentality and spatial control of waste that allows these materials to be produced since it is exported domestically and internationally, which works hand in hand with domestic residential segregation and spatial/ideological segregation transnationally.

6. Bullard 2000, Pellow 2002, Pellow and Brulle 2005.

7. These figures come from Goldstein and Madtes 2001 (see Sze 2007).

8. Horrigan and Motavalli 1997.

9. The proportion of paper recycled in the United States has increased steadily, from 18 percent in 1960 to more than 45 percent in 2000. But this has not risen fast enough to offset increased consumption, which means that the nation is still abandoning more paper to landfills and incineration than it was forty years ago (BBC News 2002). General waste recycling in the United States increased from 5 percent in 1960 to around 30 percent in 2000 (ibid.).

10. In fact, that may be why Mexico recycles more glass than the United States does. See Basel Action Network 2002b. With regard to the international movement of waste for recycling, we know that—in the wake of the passage of the Basel Convention on the Control of Transboundary Movements of Hazardous Wastes and their Disposal—many nations have engaged in what is called "sham recycling": labeling waste shipments as recyclable when they are actually hazardous and intended

for dumping in the South. "Historically, hazardous waste recycling has proven to be an environmental nightmare even in rich developed countries. For example, a full eleven percent of US Superfund priority sites slated for clean up were caused by recycling operations" (BAN 2002b).

11. BBC News 2002.

12. Friends of the Earth UK 2003.

13. "Waste Not, Want Not," 2002 (cf. Environmental News Service, 2003a).

14. Environmental News Service 2003g. For U.S. examples see Szasz 1994 and Weinberg, Pellow, and Schnaiberg 2000.

15. Environmental News Service 2005. Fly dumping is a problem in many parts of the world (see Pellow 2002).

16. BBC News 2002.

17. Lavery 2005.

18. Abbasov 2004. In January 2005, government workers removed 750 tons of garbage from Nairobi, Kenya's Wakulima market. The effort also included the slaughter of six thousand rats and removal of 70 tons of human waste from latrines in the first major cleanup of this market in thirty years. Rubbish had piled up seven feet deep in some places, and city workers used more than 42,000 gallons of water in the cleanup operation (Associated Press 2005b). On that same note, urban residents in India produce an estimated 100,000 tons of waste per day, and it is a growing problem (www.toxicslink.org).

19. "China Accuses U.S. of Dumping Illegal Waste," 1996.

20. Ibid.

21. Faison 1997.

22. Fawzi and Razaq 2004. The fact that much of the U.S. military abuse against Iraqi insurgents and prisoners in the Abu Ghraib prison scandal involved simulating pornographic behavior likely exacerbates local residents' view of similar material in the dumps.

23. See United Church of Christ 1987, Bullard 2000.

24. Cf. Getches and Pellow 2002.

25. Tangri 2003.

26. Powell 1984.

27. Tangri 2003. See also Ward 1987.

28. Montague 1998b.

29. Connett 1998.

30. Szasz 1994.

31. Tangri 2003, 66.

32. Ibid.

33. Cf. Tangri 2003, note 180.

34. Author interview with Bobby Peek, August 24, 2004.

35. Skrzycki and Warrick 2000. In this article, the authors discuss a long-awaited report by the Environmental Protection Agency that concluded for the first time that dioxin is a "human carcinogen." The report notes that dioxin emissions declined from their peak levels in the 1970s but still may pose a significant cancer threat to some people who ingest the chemical through food. "It's the Darth Vader of toxic chemicals because it affects so many systems [of the body]," said Richard Clapp, a cancer epidemiologist at Boston University's School of Public Health. "The amounts are coming down, but even small amounts are harmful."

36. Leonard 2005a.

37. I will note again that there is no need for a race-versus-class debate in environmental justice studies because environmental racism/inequality is always about both. Additionally, this framework must allow for the inequality between and within nations (see Newell 2005).

38. Other scholars and many journalists have reported on this case. What makes this presentation unique is (1) the focus on the role of activist networks in this campaign as a result of a greater level of documentation from U.S .and Haitian activist groups involved, to which I was privy; (2) a consideration of the ways in which the stakeholders framed the issue in terms of environmental injustice and inequality by race, class, and nation; and (3) the inclusion of information concerning the resolution of this case, something that other scholars were not able to discuss because the scandal was drawn out for so long (see, for example, Clapp 2001, 34).

39. The other well-known case of a wandering toxic cargo ship was that of the *Mobro 4000,* or the "garbage barge," which left New York City for North Carolina in March 1987 with three thousand tons of trash, only to return to New York City in July to eventually see its trash cargo burned in the Long Island incinerator. That was a domestic case. Earlier cases of transnational waste dumping by the United States had occurred in Mexico.

40. Montague 1987a.

41. Falsely labeled toxic waste shipments became an infamous and common method of shipping hazardous waste across national borders, particularly after the Basel agreement went into force.

42. Schwartz 2000.

43. Witness for Peace-Mid Atlantic 1998. This was a strong statement considering that by the late 1990s, the United States had shipped waste to many other parts of the world, including South Africa, Bangladesh, and India, in what Greenpeace (1998d) called "testaments to an ongoing record of environmental racism and injustice by the US."

44. Montague 1998a.

45. Bruno 1998b. This has been a common EPA practice, capping and containing the waste in place—a scientific response rather than a politically and ecologically

just response. For a study that underscores how widespread this practice is, see Lavelle and Coyle 1992.

46. Schwartz 2000.

47. Sawyer 1992.

48. Ibid.

49. Revkin 1998. The relationship between governments and corporations and their subcontractors who are often engaged in unsavory business is a dynamic that we see across many social movement struggles (for example, the antisweatshop movement and its struggle with the Jessica McClintock company, which subcontracted with smaller firms to make dresses under deplorable working conditions, thus shielding McClintock from liability) and is frustrating for activists when the former take no responsibility for their partners' actions. This problem arises in the case of PepsiCo's plastic bottle dumping in India as well.

50. The Basel Action Network, Essential Action, Greenpeace, Witness for Peace, Friends of the Earth International, and Global Response were the U.S. member organizations in the Project Return to Sender coalition. The main Haitian group in the coalition was COHPEDA (Collectif Haitian pour la Protection de l'Environnement et une Developpement Alternatif, or Collective for the Protection of the Environment and an Alternative Development), supported by the Haiti Communications Project (based in Cambridge, Massachusetts) and ROCHAD (Regroupement des Organismes Canado-Hatiens Por le Developpement, based in Canada). This was the second major coalition organized around this case. The first formed in February 1995 by Greenpeace, COHPEDA, the Haiti Communications Project, and others, calling itself the Coalition for the Return of the Toxic Waste of Gonaives.

51. Project Return to Sender 1997. This language originated in a memo sent to Project Return to Sender by Jim Puckett, executive director of the Basel Action Network, Seattle, Washington, March 20, 1998.

52. Project Return to Sender 1997.

53. Bruno 1998a.

54. Associated Press 1998.

55. Essential Action 1998.

56. Ibid.

57. Leonard 1998. Another creative and humorous approach was the title of a press release Essential Action created in response to a coalition of University of Pennsylvania student groups that organized to pressure the mayor of Philadelphia to bring the ash back. The title of that press release read: "University Students Tell Rendell to Get His Ash Home!" (Essential Action 1999 April 16).

58. Beauchamps 1998c.

59. Beauchamps 1998b.

60. COHPEDA 1998.

61. Beauchamps 1998.

62. Capek 1993.

63. Beauchamps 1998a. It is revealing to note that the concern over international waste dumping in Haiti was so great that a ban was explicitly written into the national constitution. Article 258 was passed shortly after the *Khian Sea* waste created a scandal.

64. Global Response 1998.

65. Rendell 1998.

66. Palmer 1998.

67. Piette 1998.

68. Greenpeace 1998b, emphasis added.

69. While Greenpeace's work is unparalleled and of extraordinary value, I must point out that colonialism is a cultural and political-economic system, while clean production is a largely technical approach, which reveals that many environmentalists sometimes miss the point that environmental injustice (or "toxic colonialism") concerns the exploitation of both people and their environment.

70. Leonard 1999.

71. Morel 1999.

72. Karshan 2000. Karshan was the foreign press liaison for President René Préval.

73. See LaDuke 1999, Pellow 2002.

74. Beans 2002.

75. This was similar to what happened with the infamous Shintech chemical company, which attempted to locate a PVC chemical plant in an African American community in Louisiana in the mid-1990s. Eventually activists repelled Shintech, and the plant located in Plaquemine, Louisiana, a mostly Anglo town (see Roberts and Toffolon-Weiss 2000).

76. Fiorillo and Spikol 2001.

77. Ibid.

78. Park 1924.

79. Fiorillo and Spikol 2001.

80. Ibid.

81. Gottlieb 2001, Pellow and Park 2002, Peña 1997, D. Taylor 1997.

82. O'Neill 2000.

83. Montague 1998a.

84. However, as much as sham recycling may be problematic, it is also the case that recycling itself is a sham if we believe that it represents the key to a prosperous and environmentally just future. There are simply too many limitations involved in materials recycling practices for it to hold the promise of environmental justice if production processes themselves are not altered and zero waste polices are rejected. See Gould, Weinberg, and Schnaiberg 1996; Weinberg, Pellow, and

Schnaiberg 2002; and Pellow 2002 for critiques of recycling's social, economic, and environmental impacts, particularly on working-class, poor, immigrant, and people-of-color communities.

85. Interview with the author, March 7, 2002.

86. Ibid.

87. For example, see Knight 2001, and "Japan: Philippines Case Bares Inadequacy of Waste Rules," 2000.

88. In that sense, there was a real possibility that other southern nations—and then northern nations—would follow suit, from the ground up, globally.

89. Greenpeace 1999a.

90. Ibid.

91. Basu 1999.

92. Ibid.

93. Ibid.

94. Ibid.

95. Tangri 2003, 73.

96. Ibid.

97. The irony is that, dioxins aside, most incineration technology in Europe is comparatively safer and cleaner than the technology the same companies are selling to southern nations, which is also an example of environmental inequality.

98. This is a common tactic among interest groups opposing progressive legislation in various parts of the world. The U.S. government's trade representative and the American Electronics Association did the same in Europe in response to efforts to pass the WEEE and RoHS Directives for the EU.

99. Author interview with Von Hernandez, February 24, 2004.

100. Ibid.

101. Ibid.

102. Leonard 2005a.

103. Author interview with Hernandez.

104. Ibid.

105. When activists take their arguments to governing bodies, this is called "venue shopping" (see Baumgartner and Jones 1991).

106. Author interview with Hernandez.

107. Author interview with Jorge Emmanuel, March 16, 2004.

108. This dynamic is called a media shift; that is, once one medium of waste management is blocked or heavily regulated, polluters and waste managers often shift the waste to another medium—air, land, or water—rather than pursuing actual toxics reduction and ecologically sustainable policies. See Pellow 2002.

109. Author interview with Hernandez.

110. For an analysis of this point as it applied to the civil rights movement, see Morris 1984.

111. Author interview with Hernandez.

112. GAIA 2001.

113. Author interview with Francis de la Cruz, February 4, 2004.

114. Ibid. Repression against human rights and environmental justice activists is ongoing in the Philippines. In June 2006, Markus Bangit, a tribal leader and activist with the Cordillera Peoples Alliance, a group fighting against World Bank–supported dam projects, was assassinated by a hooded gunman. This was the latest in a series of political killings of environmental activists in that nation. Letter to the author from Aviva Imhof, campaigns director, International Rivers Network, July 2006.

115. Author interview with Jorge Emmanuel, March 16, 2004.

116. Author interview with de la Cruz.

117. Ibid.

118. Author interview with Emmanuel.

119. Ibid.

120. For local coverage of this case see: "Pepsi 'Dumping' Plastic Waste," 1994; "Pepsi Accused of Flouting Trade Law," 1994; "Pepsi's Dumping Strategy Raises Public Outcry," 1994; "'Pepsi Failed to Fulfill Promise,'" 1994.

121. Leonard 1994.

122. Ibid. Leonard writes that the "anti-Pepsi campaign is part of a larger 'Swadeshi' movement urging the boycott of many foreign goods in the country. The term 'Swadeshi' has been borrowed from the movement launched by Mahatma Gandhi against British-made goods during India's independence struggle. At that time, Gandhi called upon the British to 'quit India'; now activists are calling on multinational corporations—including Pepsi—to quit India." (7)

123. Ibid. 10.

124. Ibid.

125. While the Public Interest Research Group movement began in the United States with Ralph Nader in the 1970s, this particular PIRG describes itself as a campaign and research nongovernmental organization involved in structural adjustment, debt, transnational corporations, and developmental issues.

126. Global Response 1994.

127. Gottlieb 2001.

128. Global Response Action 1994.

129. Ibid. The figure of thirty cents a day comes from Greenpeace 1994. In the same document, Greenpeace also reported that PepsiCo's CEO Wayne Calloway enjoyed a salary and bonuses of more than $2 million annually.

130. Leonard 1994, 8.

131. See Hossfeld 1990, Pellow and Park 2002, Zavella 1987.

132. Global Response Action 1994.

133. Jaitly and Singh 1994. Details regarding the protests against Pepsi's neon signs near the tomb were contained in PIRG 1994b. The tomb, built in 1753, is believed to be India's last structure built in the Mughal architectural style.

134. Greenpeace 1994.

135. Global Response Action 1994.

136. Ibid.

137. Singh 1994a.

138. Public Interest Research Group 1994b, 1. This PIRG report described an extreme case of trade liberalization in which the Indian government agreed to import cow dung from the Netherlands. PIRG claimed that the reason the Dutch were exporting the waste was because of the "chemical residues which leak into the ground water and thereby pollute it." The company that planned to ship the waste was appropriately named *Envirodung*. The whole affair blew up and "was abandoned after widespread protest in the press and on the streets of India. Farmers and milkmen agitated by the news launched articulate and malodorous protests such as planning to dump 50 tons of 'swadeshi' dung outside the Indian Parliament. The Government was forced to clarify that it would not permit the import of animal dung" (19).

139. Mokhiber 1994. In 1996, Pepsi finally consented to sell its equity stakes invested in Burma, admitting publicly that it was yielding to pressure by human rights activists (Freeman 1996). The new CEO who made the announcement (Wayne Calloway's successor) was formerly in charge of Pepsi's beverage division when the company pulled its investments out of South Africa in 1985, during the global antiapartheid movement's successful campaign to push companies to divest from that nation's racist government.

140. Letter from Brad Shaw, manager, public relations, Pepsico Foods and Beverages International, October 24, 1994. This form letter was sent to persons who participated in the letter-writing campaigns initiated by Global Response and Greenpeace regarding the charge that Pepsi was responsible for dumping plastic waste in India. I obtained copies through permission from Global Response, as the staff had been sent copies by letter recipients.

141. "Waste Dumping Grounds of the World," 2005.

142. Letter from Brad Shaw, 1994.

143. Devraj 2003.

144. "HC Stays Order on Pepsico License Cancellation," 2004. Perhaps predictably, Pepsi responded to these general concerns by launching a new venture with Aquafina, the drinking water company, to distribute that product to consumers across India. See "Pepsi Forays into New Business with Aquafina Bulk Water," 2003 and Majumder 2006.

145. "Greenpeace Indicts Pepsi for Dumping Waste in India," 1994. Note that the export of plastic waste that begins its journey through collections in recycling programs in the United States parallels exactly the process through which most of the electronic waste that ends up in Asia and Africa finds its way there (see chapter 6).

146. Greenpeace 1992.

147. For analyses of the political economy of recycling programs in the United States, see Gould, Weinberg, and Schnaiberg 1996; Weinberg, Pellow, and Schnaiberg 2000.

148. Author interview with Nityanand Jayaraman, March 5, 2002.

149. Ibid.

150. Department of Government and Foreign Trade 2001.

151. Leonard 1993. UBINIG is the abbreviation of its Bengali name, Unnayan Bikalper Nitinirdharoni Gobeshona (Policy Research for Development Alternatives).

152. See Greenpeace 1998c. *Delayed dumping* is a term Greenpeace activists have used to describe practices involving the transfer or donation of technologies that are one step away from the end of life, which means they will soon have to be dumped by the new host. Hernandez (2002) used this term to describe the way that computers are dumped on the South when companies donate them to schools when they are on their last leg.

153. Leonard 2005a.

154. Calonzo 2004.

155. Multinationals Resource Center and Health Care Without Harm 1999.

156. Health Care Without Harm. N.D. How to Shut Down an Incinerator. Washington, D.C., p. 13.

157. Leonard 2005a.

Chapter 5

1. Ehrlich 1989, xv.

2. Ramasamy 2002.

3. Weir 1987; se also Frey 1995, 152.

4. See Eddleston et al. 2002, Liden 1995. Liden challenges the claim that pesticides use leads to higher yields. The author considers how modernization has pushed Malaysia into a hyperdevelopment phase. See also "Malaysia: Project MY60: Cooperation with Consumers Association of Penang" (www.helvetas.ch/english/projects/emalaysia_my60.html).

5. Cesar Chavez was a leader of the United Farm Workers for many years and was responsible for connecting U.S. consumer behavior to the toxic working conditions of migrant laborers who picked the produce that ends up on the dining room and kitchen tables of every family in that nation. Rachel Carson is the famous

scientist and author of *Silent Spring* (1962/2002), a book credited with starting the modern U.S. environmental movement. In that book, Carson presented evidence that pesticides and other chemicals were wreaking unprecedented havoc on ecosystems and human bodies around the world.

6. Throughout history, many human populations have also been described and viewed as invasive alien species. These may include immigrants, people of color, and various marginalized "others" (see Santa Ana 2002).

7. Including animals such as rabbits and birds that sustain themselves on plants and insects that pesticides affect. See European Commission's Scientific Committee of Plants 2002 (see Madeley 2002).

8. Widespread use of pesticides for the purpose of suicide has been documented in India, Sri Lanka, Malaysia, Trinidad, Costa Rica, and Samoa (see Dinham 1993, 1996). Regardless of whether poverty and debt were the causes, medical studies and public health reports reveal that suicide by pesticide is common in many southern nations (see Eddleston 2000, Eddleston et al. 2002, Sri Lankan Ministry of Health 1997, World Health Organization 1990).

9. Delaplane 2000.

10. Steingraber 1997, Wargo 1998.

11. The majority of pesticides produced in the North are actually traded or sold among northern nations, but most of those that are banned because they are deemed most hazardous are sent southward.

12. Steingraber 1997, 166; Weir and Schapiro 1981; WHO 1992.

13. Environment News Service 2003f. Scientific studies lend support to this statistic (see Frey 1995, Jeyaratnam 1990, World Health Organization 1990).

14. *Toxic Trail* 2001.

15. Pesticide Action Network North America 2003.

16. U.S. government spot checks find that approximately 10 percent of all imported food is contaminated with illegal levels of pesticides. The Food and Drug Administration (FDA) reports that nearly half of the green coffee beans imported to the United States are contaminated with pesticides that have been banned in the United States. At least twenty pesticides, potential carcinogens, are undetectable with FDA tests used to find residues in food (Greenpeace 1999b). See also Pennycook, Diamand, Watterson, and Howard 2004.

17. *Toxic Trail* 2001.

18. Excerpted from the United Nations Economic and Social Council 58th session, Item 7 on the provisional agenda, Section E.

19. United Nations Population Fund 2001, 38; London et al. 2002.

20. Wesseling et al. 1997.

21. Fernandez 2002.

22. M. Goldsmith 1989; Quintero-Somaini et al. 2004, 38–49.

23. A. Garcia 2003, Dewailly et al. 1994, Wolff et al. 1993.

24. Pesticide Action Network Asia and the Pacific 2003d. As another study concluded, "Organochlorine insecticides [including aldrin and DDT] . . . act as antagonists to pregnancy" (Saxena et al. 1981).

25. Environmental News Service 2003f; Leslie et al. 2002; Shiva 1988, 1991.

26. Pesticide Action Network Asia and the Pacific 2003c,d.

27. International Labor Organization 1994.

28. Ibid. 1996.

29. Jeyaratnam, Lun, and Phoon 1987; Moses 1993.

30. Pesticide Action Network Asia and the Pacific 2003, 5.

31. See Dennis and Pallota 2001; Reeves, Schafer, and Katten 1999.

32. Bray 1994; Frey 1995, 158. See also Dinham 1993; Garcia 2003; London et al. 2002; WHO 1990, 1992.

33. Tenaganita and PANAP. 2002. Poisoned and Silenced: A Study of Pesticide Poisoning in the Plantation. March.

34. Clapp 2001; Epstein 1979; Steingraber 1997; Toxics Link India 2002.

35. Toxics Link India 2002.

36. Barnet and Muller 1974; Gould, Weinberg, and Schnaiberg 1996; Smith, Sonnenfeld, and Pellow 2006a; Starr 2000.

37. Tenaganita and PANAP 2002. Poisoned and Silenced: A Study of Pesticide Poisoning in the Plantation. March.

38. Tenaganita and PANAP 2002. Lack of translations concerning hazardous materials is a common problem in many nations where immigrant farmworkers labor (see Eddleston et al. 2002).

39. Philo and Miller 2001, Schaeffer 2003.

40. Frey 1995, 153.

41. Pesticide Action Network Updates Service 2002. Excerpted from a paper originally published in the *International Journal of Occupational and Environmental Health* (Vol. 7, No. 4).

42. Environment News Service 1998.

43. U.S. Newswire 2001.

44. Madeley 2002, 7.

45. Larsen 1998.

47. Madeley 2002, 6.

47. Lemos and Jayaraman 2002.

48. Frey 1995, 155.

49. Madeley 2002.

50. Awuonda 2001.

51. Global Pesticide Campaigner 2003.

52. "More Were Exposed to Agent Orange," 2004.

53. Doyle 2004, 78.

54. Global Pesticide Campaigner 2003. See also the United Nations Economic and Social Council 58th session, Item 7 on the provisional agenda Section E.

55. Russell 2001, 36 , chapter 2.

56. Druley 1997.

57. Schafer 2001.

58. Hough 2000.

59. Clapp 2001.

60. "Eco-Reports: Pesticides in the English Speaking Caribbean," n.d.

61. Ministry of Health 1992.

62. Ibid.

63. Global Response 1992b.

64. Gladstone 1992c. Note that the environmental laws within the United States make such accountability much clearer, as is the case with brownfields and the "joint and several liability" requirement that all previous owners of abandoned toxic sites are liable for future cleanups.

65. Global Response 1992b.

66. Gladstone 1991.

67. Gladstone 1992b.

68. ReEarth's spokesperson was Sam Duncombe. See Gladstone 1992a.

69. Leonard 1991.

70. See Clapp 2001.

71. Braynen 1992. There were also letters from the Bahamian embassy to Ann Leonard at Greenpeace International on November 19, 1991, and August 3, 1992.

72. Campbell 1992.

73. Representatives from the Owens-Illinois company would also later ask Arbury for assistance in finding local recyclers for some of the nonhazardous materials on site.

74. Leonard 1992.

75. Global Response 1992a.

76. Hoff 1992.

77. reEarth 1992.

78. Mrs. Sam Dumcombe, spokesperson from reEarth, in an undated letter to Malcolm Campbell of Global Response.

79. reEarth 1992.

80. Gladstone 1992a.

81. Towles 1993.

82. "Toxic Chemicals Leaving Abaco," 1994.

83. "Toxic Waste Leaves Abaco," 1995.

84. "Toxic Chemicals Leaving Abaco," 1994.

85. reEarth 1992.

86. Thompson n.d.

87. UNEP n.d. According to a report titled *Environmental Agenda for the 1990s*, in the eastern Caribbean countries of Antigua and Barbuda, Dominica, Grenada, St. Kitts and Nevis, St. Lucia, and St. Vincent and the Grenadines, there is little domestic capacity for detection and monitoring of what are believed to be significant pesticide levels in drinking water, farm produce, soil, or human tissue.

88. Lemos and Jayaraman 2002, Global Pesticide Campaigner 2002.

89. Global Pesticide Campaigner 2002. Partners in the Africa Stockpiles Program include the African Development Bank; the African Union; the Basel Convention Secretariat, Canada; CropLife International; the European Union; the Food and Agricultural Organization; the French Republic; the Global Environment Facility, Japan; the Kingdom of Belgium; the Kingdom of Norway; the Kingdom of Sweden; the New Partnership for Africa's Development; the Pesticide Action Network Africa; the Pesticide Action Network UK; the United Nations Economic Commission for Africa; the United Nations Environment Program, the United Nations Industrial Development Organization; the United Nations Institute for Training and Research; the World Bank; the World Health Organization; and the World Wildlife Fund.

90. Barclay and Stegall 1992.

91. Brough 2001.

92. Environment News Service 2003d.

93. Conversation with anonymous activist, fall 2005.

94. Lemos and Jayaraman 2002.

95. Pesticide Action Network North America n.d.

96. It is critical that we understand why so many African nations are fiscally devastated. Africa is perhaps the richest place on earth with regard to its natural resources. So how can it be the site of such extreme poverty? As Naomi Klein (2005) has so aptly written: "This is what keeps Africa poor: not a lack of political will but the tremendous profitability of the current arrangement. Sub-Saharan Africa, the poorest place on earth, is also its most profitable investment destination. It offers, according to the World Bank's 2003 Global Development Finance report, 'the highest returns on foreign direct investment of any region in the world'. Africa is poor because its investors and its creditors are so unspeakably rich."

97. Environment News Service 2004a, Redefining Progress 2004.

98. The neocolonial nature of dependent relations between many southern and northern nations plays out every day in Mozambique. Foreign donors manage public goods there, creating, according to one observer, an "internationalized"

Mozambican state: "Parameters are first set in Paris among a group of donors, and only afterwards is the national budget discussed in Maputo. Issues of national importance are first handled via satellite between Washington D.C. and Maputo before they circulate to the public in Mozambique" (Sogge n.d.).

99. As an indicator of how repression against civil society and free expression remains a serious problem in Mozambique, Carlos Cardoso, an investigative journalist for the muckraking newspaper *Metical* (which broke the Danida case), was assassinated after uncovering corruption between well-placed elected officials and the Bank of Mozambique in the late 1990s (Wines 2006, Hanlon 2001).

100. Puckett 1998.

101. Ibid.

102. Neilsen 1999.

103. This "English-only" model of environmental impact assessments is something other communities of color have had to contend with. In a famous example, El Pueblo para Aire y Agua Limpio (People for Clean Air and Water) won a landmark legal case against the firm Waste Management when the environmental impact statement for a new landfill in Kettleman City, California, was written only in English, despite the fact that a significant percentage of residents spoke only Spanish (see Cole and Foster 2001).

104. Author interview with Anabela Lemos, February 5, 2004.

105. Puckett 1998.

106. Mangwiro 1999.

107. Greenpeace International 1998, Puckett 1998.

108. Lowe 2003.

109. Ibid.

110. Ibid.

111. Author interview with Lemos.

112. Basel Action Network 1998.

113. Lowe 2003.

114. Author interview with Lemos.

115. Puckett 1998.

116. Neilsen 1999.

117. Africa News Service 1998.

118. Coalition Press Release 2000.

119. Author interview with Lemos.

120. Ibid.

121. Environmental News Service 2000.

122. Puckett 1998.

123. Environmental News Service 2000.

124. Author interview with Lemos.

125. Lemos 2003.

126. Mediafax 2002.

127. Author interview with Lemos.

128. Central Intelligence Agency 2005.

129. Winant 2001.

130. Dunlap and Marshall 2006.

131. Lemos and Jayaraman 2002.

132. Greenpeace USA n.d.

133. Pesticide Action Network Asia Pacific 2003b.

134. In October 2000, AstraZeneca and Novartis merged to create Syngenta, the world's largest agro corporation.

135. Rengam and Mourin 2002.

136. Author interview with Mageswari Sangaralingam, fall 2004.

137. Environmental News Service 2003c.

138. Guinier and Torres 2002.

139. Ramasamy 2002.

140. Pesticide Action Network Asia Pacific 2003a.

141. Ransom n.d. Agenda 21 is a comprehensive plan of action to be taken globally, nationally, and locally by organizations of the United Nations system, governments, and major groups in every area in which humans have an impact on the environment. More than 178 governments adopted the plan at the United Nations Conference on Environment and Development held in Rio de Janeiro, Brazil, in June 1992.

Chapter 6

1. Agarwal and Wankhade 2006.

2. For example, more than two-thirds of adults in the United States actively use computers and the Internet (U.S. Department of Commerce 2002).

3. Pellow and Park 2002.

4. The doubling of the microchip speed is known as "Moore's law," named after Intel Corporation founder Gordon Moore, who in 1965 predicted this trajectory for that technology's development.

5. Silicon Valley Toxics Coalition 2001, 2; Slade 2006.

6. See also Grossman 2006.

7. Reuther 2002.

8. Osborn 2002.

9. Basel Action Network and Silicon Valley Toxics Coalition 2002, 1.

10. Silicon Valley Toxics Coalition 2001, 3.

11. The 1985 statistic comes from Toxics Link 2003. The 2006 statistic comes from Leo 2006.

12. Environmental News Service 2003b.

13. Silicon Valley Toxics Coalition 2001, 7.

14. Environmental News Service 2004b. The first "Eco-PC" was produced by Siemens Nixford in 1993, but sales were lower than expected and "even environmental organizations chose cheaper conventional models."

15. Sample 2004.

16. Gould, Schnaiberg, and Weinberg 1996; Schnaiberg 1980; Weinberg, Pellow, and Schnaiberg 2000.

17. "Cold Water Poured on IT's Environmental Pluses," 2003. The full report is in Kuhndt et al. 2003. The material reality of our reliance on transport in a "virtual economy" is summed up well in a slogan on the back of many commercial trucks in New England: "If you got it, a truck brought it."

18. Bullard 2000.

19. Basel Action Network and Silicon Valley Toxics Coalition 2002; GrassRoots Recycling Network 2003a.

20. Author interview with Jim Puckett, March 5, 2002.

21. Gough 2002.

22. Basel Action Network and Silicon Valley Toxics Coalition 2002, 29.

23. Moyers 1990.

24. Author interview with Ted Smith, spring 2002. The Common Sense Initiative (CSI) was a multistakeholder effort that the U.S. EPA convened in the 1990s as a way to placate industry and labor unions, which felt that environmental regulations were too stringent, and to placate environmentalists and environmental justice activists, who believed the regulations were too weak. The CSI aim was to produce "cleaner, smarter, cheaper" environmental management and regulatory frameworks. Unfortunately the effort failed when industry participants used it as a way to circumvent regulations and environmentalists' push for clean production was all but ignored.

25. Public Interest Research Group 1994a, 5.

26. Capek 1993.

27. Computer TakeBack Campaign 2003.

28. GrassRoots Recycling Network 2003a.

29. Vidal 2004.

30. Goodman 2003.

31. Asia Pacific n.d.

32. Author interview with Puckett.
33. Japan Economic Newswire Plus 1994.
34. Silicon Valley Toxics Coalition 2001, 18.
35. Author interview with Puckett.
36. Author interview with Smith.
37. Basel Action Network and Silicon Valley Toxics Coalition 2002.
38. Ibid. 22.
39. BBC News 2002.
40. Brigden, Labunska, Santillo, and Allsopp 2005.
41. Author interview with Smith.
42. Black 2004.
43. Glueck 2003.
44. Iles 2004.
45. Xing 2003.
46. For additional information on the situation in China, see Leong and Pandita 2006. For more on e-waste in China and Asia more generally, see Iles 2004.
47. The Times of India Online 2004.
48. Quoted in Asia Pacific n.d.
49. Kamdar 2004.
50. Toxics Link India 2003, 5.
51. Toxics Link 2003, 6.
52. Wankhade 2004.
53. Ibid.
54. Ibid.
55. Agarwal and Wankhade 2006, Ghosh 2006, Pandita 2006.
56. Agence France-Press 2003.
57. Hernandez 2002.
58. BBC News 2003a, June 6.
59. Ted Smith of the Silicon Valley Toxics Coalition stated that in his experience, CAI is very careful to test every computer before export (communication to author, June 2006).
60. BBC News 2003a, June 6.
61. Kanyegirire 2003.
62. Vidal 2004.
63. Basel Action Network 2005.
64. Silicon Valley Toxics Coalition 2001, 18.

65. Garcia and Simpson 2006.

66. Estrada 2004.

67. Basel Action Network and Silicon Valley Toxics Coalition 2002, 8.

68. Reuther 2002.

69. www.texasenvironment.org, emphasis added.

70. GrassRoots Recycling Network 2003b, emphasis added. GRRN no longer works on e-waste issues.

71. Silicon Valley Toxics Coalition Web site: www.svtc.org, emphasis added.

72. Basel Action Network Web site: www.ban.org, emphasis added.

73. Author interview with Smith.

74. Computer TakeBack Campaign 2003, emphasis added.

75. Toxics Link 2003, 2, emphasis added.

76. Wenling Tu communication with the author, August 7, 2006.

77. Basel Action Network and Silicon Valley Toxics Coalition 2002.

78. IBM had recently sold its brand to Lenovo, a Chinese company, so IBM no longer sells computers to consumers. "So they want the computer industry to pay for their share of their historic waste, which is what would happen if the system were funded by retail sales, since they are no longer selling consumer products and their mountains of e-waste that have accumulated over the many years that they were the leader in selling computers to consumers would have to be paid by others under their approach" (Smith 2005).

79. Associated Press 2003a.

80. Schatz 2003.

81. Computer TakeBack Campaign 2003. Sheila Davis is now the executive director of SVTC, and Smith is a senior campaign strategist.

82. Anonymous UNICOR worker quoted in Dayaneni and Doucette 2005, 14.

83. Ibid.

84. Computer TakeBack Campaign 2003.

85. Ibid.

86. Associated Press 2003b.

87. Texas Campaign for the Environment 2003a.

88. Ibid.

89. "Dell Cancels Contract for Inmate Labor," 2003.

90. Computer TakeBack Campaign 2002.

91. Texas Campaign for the Environment 2003b.

92. Raphael and Smith 2006.

93. Flynn 2004.

94. Bartholomew 2004.

95. Recycling Today 2004.

96. Sangaralingam 2004a.

97. Yatim 2004.

98. Ibid.

99. Sangaralingam 2004b.

100. Author interview with Mageswari Sangaralingam, February 26, 2004.

101. Ibid. In 2005, Citiraya was rocked with a scandal involving former employees of Advanced Micro Devices, 3M, Seagate, and STMicroelectronics. The allegation was that Citiraya executives bribed former employees of these companies to resell microchips on the open market despite the fact that they were destined for destruction. Several persons were sentenced to jail, one of whom—a Citiraya executive—fled to evade authorities. In the wake of this scandal, Citiraya declared bankruptcy (see EE times on-line, http://www.eetimes.com/news/semi/showArticle.jhtml?articleID=174403245 December 26).

102. M. Williams 2006.

103. Ibid.

104. "Computer Giant HP Mute over Toxin Use," 2005.

105. Sherriff 2005.

106. Greenpeace USA 2006a.

107. May 2006.

108. "Greenpeace Launches Campaign against E-Waste," 2005.

109. Greenpeace and Basel Action Network n.d.

110. Basel Action Network and Silicon Valley Toxics Coalition 2002, 3.

111. Basel Action Network 2002a. Keep in mind that China had already banned e-waste imports as of 2000 (Reuters News Service 2005, March 31).

112. Leong and Pandita 2006, 55.

113. Author interview with Smith.

114. Wankhade 2004.

115. Baxter 1999, 7.

116. None of the legislation in the United States prevents exports. While just six U.S. states (California, Maine, Minnesota, Washington, Maryland, and Massachusetts) have passed e-waste legislation, more than half the states have introduced (but not yet passed) similar legislation (Ted Smith correspondence with the author, June 15, 2006).

117. For example, recently many outgoing computer shipments from the United Kingdom have been confirmed as e-waste despite being labeled "product," thus violating the Basel Ban (personal communication with Ted Smith, June 2006).

118. Ibid.

119. Basel Action Network 2005.

120. Stiffler 2004.

121. The California law bans some forms of these chemicals but not the ones used in electronics (known as deca PBDEs). Ted Smith correspondence with the author, June 15, 2006.

122. Knight 2000. The *Inside EPA* newsletter was published September 15, 2000.

123. Ibid.

124. Ibid. After Sony made the presentation, as far as activist Ted Smith is aware, the surveillance plan "was never implemented. We think that the publicity helped to deter them from carrying this forward—they were mighty embarrassed by this" (Ted Smith correspondence with the author, June 15, 2006).

125. For example, see Veldhuizen and Sippel 1994, 11.

126. Byster 2001.

127. Computer TakeBack Campaign Platform, March 3–5, 2001, Washington, D.C.

128. Knight 2000.

129. Cahalan 2003.

130. Dean 2003.

131. Pellow 2001.

132. Author interview with Smith.

Chapter 7

1. Beck 1995, 27.

2. Guinier and Torres 2002; Roediger 1999; Kivel 2002; Wise 2001, 2005.

3. Feagin, Vera, and Batur 2001, 19.

4. Ibid., 19–23.

5. Keck and Sikkink 1998, 200.

6. This is in reference to language contained in the U.S. Supreme Court's decision in the *Brown* v. *Board of Education* case to strike down racial segregation as unconstitutional. Many forms of social change do in fact occur with all deliberate speed, so most seasoned activists maintain a long view regarding movement goals and time lines.

7. O. Williams 2000.

8. Beck 1995.

9. Florini 2000, Keck and Sikkink 1998, Roberts and Grimes 2002.

10. Tarrow 1998, 194–195.

11. Chua 2003; Lipsitz 2001, 17.

12. Lipsitz 2001.

13. For excellent writings about racial states, see Goldberg 2002 and James 1996.

14. Bonilla-Silva 2003, Winant 2001.

15. Gilroy 2000, Darder and Torres 2004. For excellent critiques of postracial scholarship see Barlow 2003, Bonilla-Silva 2001, and Guinier and Torres 2002.

16. "At present there is no well-developed theory of class in all of its aspects, which remains perhaps the single biggest challenge facing the social sciences. Indeed, failure to advance in this area can be seen as symptomatic of the general stagnation of the social sciences over much of the twentieth century" (Foster 2006). For an excellent critique of scholars who have retreated from a discussion of both race and class, see Roediger 2006.

17. Boggs 2000.

18. As one author observed, "Forty-seven of the top one hundred economies of the world are actually transnational corporations; 70 percent of global trade is controlled by just five hundred corporations; and a mere 1 percent of the TNCs on this planet own half the stock of foreign direct investment" (Clarke 1996, 298).

19. Conversation with Heeten Kalan, September 2005.

20. Some of the many grievances include the fact that large environmental organizations (1) have negotiated "debt-for-nature" swaps that have displaced indigenous peoples from their land; (2) have supported the closure of particular firms on the grounds of pollution control without attention to the employment consequences for workers; (3) have welcomed to their boards of directors representatives from TNCs known for perpetrating environmental injustices; (4) have dominated the funding available from philanthropic sources; and (5) have dismal records as far as the racial and ethnic diversity of their membership, staff, and boards of directors is concerned (Alston 1991).

21. Author interview with Heeten Kalan, December 23, 2003.

22. Author interview with Anabela Lemos, February 5, 2004.

23. Ibid.

24. See, for example, the case of the struggle against the Shintech corporation in Louisiana during the mid-1990s, when the company was prevented from locating a PVC plant in the majority African American parish of St. James and eventually sited the plant in the mostly white town of Plaquemine, Louisiana (Roberts and Toffolon-Weiss 2001).

25. Many U.S. environmental justice movement groups decided to recommit their focus on local grassroots organizing after the EJ Summit II in 2002 failed to produce a coherent national EJ agenda. Local action is critical and the foundation of the movement, but will remain limited if transnational networks are not also a core part of this effort.

26. For additional critiques concerning the U.S. environmental justice movement, see Pellow and Brulle 2005.

27. Sachs, 1995, 45.

28. Barlow 2003. There is an emerging critique by scholars in postcolonial, feminist, and ethnic studies concerning human rights law and discourse because of

their reliance on a universalist construct of rights. That is, by universalizing this idea of rights, the UN tends to erase and subjugate the great diversity of cultures and contexts that give meaning to peoples' life experiences and their own conceptions of justice—something that colonialism did very well also. For an excellent exploration of these issues, see Hua 2006. However, the problem with this critique is that by celebrating the great diversity of human cultures and experiences and by fixating on whether or not discourses reproduce shadows of colonialism, one may ultimately contribute to the disempowerment of peoples by enabling an endless fragmentation of experiences that dominant social forces could capitalize on by arguing "there are no common standards of justice." Human rights discourses in certain hands can be used against otherwise disenfranchised persons, but they can also be used in favor of progressive politics and justice for marginal groups. It is not a standard that is inherently better than what existed before, but it can and has offered critical advances in many contexts. I embrace these tensions.

29. Advocates for Environmental Human Rights 2005, 1.

30. Anstead 2004; Cahn 2004; Kucinskaite 2002; Plese 2002, 2004; Szendrey 2002.

31. Conversation with the author, ERRC offices, Budapest, Hungary, October 2005.

32. Global Response 2005.

33. Wood 2006.

34. Global Response 2005.

35. Global Response 2006.

36. See the appendix: Principles of Environmental Justice.

37. Leonard 2005a.

38. Bullard 2005a, 25.

39. Author's interview with Sathinath Sarangi, February 2003.

References

Abbasov, Idrak. 2004. "Baku's Rubbish Dump from Hell." *Environment News Service,* August 24.

Adeola, Francis. 2000. "Cross-National Environmental Justice and Human Rights Issues—A Review of Evidence in the Developing World." *American Behavioral Scientist* 43:686–706.

Advocates for Environmental Human Rights. 2005. *Environmental Justice and Human Rights in the United States.* New Orleans: Advocates for Environmental Human Rights.

Africa News Service. 1998. "SA Dumping Plan 'Trashed' by NYC." January 8.

Agarwal, Ravi, and Kishore Wankhade. 2006. "High-Tech Heaps, Forsaken Lives: E-Waste in Delhi." In Ted Smith, David Sonnenfeld, and David N. Pellow (Eds.), *Challenging the Chip: Labor Rights and Environmental Justice in the Global Electronics Industry.* Philadelphia: Temple University Press.

Agbola, Tunde, and Moruf Alabi. 2003. "Political Economy of Petroleum Resources Development, Environmental Injustice and Selective Victimization: A Case Study of the Niger Delta Region of Nigeria." In Julian Agyeman, Robert Bullard, and Bob Evans (Eds.), *Just Sustainabilities: Development in an Unequal World.* Cambridge, Mass.: The MIT Press.

Agence France-Press. 2003. "India Is Fast Becoming a Dumping Ground for Electronic Waste." June 17.

Agyeman, Julian. 2005. *Sustainable Communities and the Challenge of Environmental Justice.* New York: New York University Press.

Agyeman, Julian, Robert Bullard, and Bob Evans (Eds.). 2003. *Just Sustainabilities: Development in an Unequal World.* Cambridge, Mass.: The MIT Press.

Alexander, Jeffrey C. 1990. "Between Progress and Apocalypse: Social Theory and the Dream of Reason in the Twentieth Century." In Jeffrey Alexander and Piotor Sztompka (Eds.), *Rethinking Progress: Movements, Forces, and Ideas at the End of the 20th Century.* Boston: Unwin Hyman.

Alger, Chadwick. 1997. "Transnational Social Movements, World Politics, and Global Governance." In Jackie Smith, Charles Chatfield, and Ron Pagnucco (Eds.),

Transnational Social Movements and Global Politics: Solidarity Beyond the State. Syracuse, N.Y.: Syracuse University Press.

Almaguer, Tomas. 1994. *Racial Fault Lines: The Historical Origins of White Supremacy in California.* Berkeley: University of California Press.

Alston, Dana. 1990. *We Speak for Ourselves.* Washington, D.C.: Panos Institute.

Alston, Dana. 1991. "Moving Beyond the Barriers." Speech delivered at the First National People of Color Environmental Leadership Summit, Washington, D.C., October.

Alston, Dana, and Nicole Brown. 1993. "Global Threats to People of Color." In Robert Bullard (Ed.), *Confronting Environmental Racism: Voices from the Grassroots.* Boston: South End Press.

Althaus, Dudley, and Edward Hegstrom. 2005. "Mexicans taken Aback by Racial Stir over Stamps." *Houston Chronicle,* July 1.

Alvarez, Robert. 2006. Presentation on *Mangoes, Chiles and Truckers: The Business of Transnationalism,* University of California, San Diego, March 8.

Amin, Samir. 1989. *Eurocentrism.* New York: Monthly Review Press.

Anderson, Carol. 2003. *Eyes Off the Prize: The United Nations and the African American Struggle for Human Rights, 1944–1955.* Cambridge: Cambridge University Press.

Anderton, D., A. Anderson, J. Oakes, and M. Fraser. 1994. "Environmental Equity: The Demographics of Dumping." *Demography* 31:229–248.

Anstead, Alan. 2004. "Litigating Discrimination in Access to Health Care." *Roma Rights: Quarterly Journal of the European Roma Rights Center,* no. 2: 75–77.

Appadurai, Arjun. 1996. *Modernity at Large: Cultural Dimensions of Globalization.* Minneapolis: University of Minnesota Press.

Appiah, Anthony. 2006. "Toward a New Cosmopolitanism." *New York Times Magazine,* January 1.

Asia Pacific. N.d. "India: Fears That IT Hub Becoming Electronic Waste Dump." http://www.abc.net.au/ra/asiapac/programs/s886414.htm. Accessed July 29, 2003.

Associated Press. 1998. "Haiti-Toxic Waste." March 13.

Associated Press. 2003a. "Computer Makers Hammered for Their Poor Handling of 'E-Waste.'" Houston Chronicle.com. January 10.

Associated Press. 2003b. "Environmentalists Fault Dell on Recycling of PCs." Las Vegas, January 10.

Associated Press. 2005a. "Australian Racial Violence Spills Over into 2nd Night." December 13.

Associated Press. 2005b. "Kenya: 750 Tons of Garbage Removed from Market." *New York Times,* January 5, p. A6.

Awuonda, Moussa. 2001. African News Service. "Pesticide Poisoning Pose Concern to Third World." June 11.

Bacon, David. 2004. "Britain's War over Managed Migration." *Z Magazine,* September, 45–49.

Bakan, Joel. 2004. *The Corporation: The Pathological Pursuit of Profit and Power.* New York: Free Press.

Barclay, Bill, and John Stegall. 1992. "Obsolete Pesticides Crisis." *Global Pesticide Campaigner* 2(1): 4–5.

Bartholomew, Doug. 2004. "E-Commentary: Computer Makers Tackle E-Waste." IndustryWeek.com, January 1.

Barlow, Andrew. 2003. *Between Fear and Hope: Globalization and Race in the United States.* New York: Rowman and Littlefield.

Barnet, Richard, and Roland E. Muller. 1974. *Global Reach: The Power of Multinational Corporations.* New York: Simon and Schuster.

Basel Action Network. 1998. "Danish Development Project Encouraging Toxic Waste Trade into Mozambique?" Press release, Copenhagen, October 5.

Basel Action Network. 2002a. "Toxic Trade Watchdog Calls US EPA's Proposal on Electronic Waste Illegal and an Affront to Environmental Justice." Press release, September 13.

Basel Action Network. 2002b. "Hazardous Waste Recycling: No Justification for Toxic Trade." Briefing paper no. 7, March.

Basel Action Network. 2005. *The Digital Dump: Exporting Re-Use and Abuse to Africa.* October 24.

Basel Action Network and Silicon Valley Toxics Coalition. 2002. *Exporting Harm: The High-Tech Trashing of Asia.* Seattle: Basel Action Network.

Basu, Radha. 1999. "Philippines Landmark Law to Throw Out Waste Incineration." *Inter Press Service,* May 7.

Baumgartner, Frank, and Bryan Jones. 1991. "Agenda Dynamics and Political Subsystems." *Journal of Politics* 53(4):1044–1074.

Baxter, Kevin. 1999. "Environmental Groups Urge Tougher Recycling Rules." *American Metal Market,* May 21, pp. 12–13.

BBC News. 2002. "China: Hi Tech Toxics." http://news.bbc.co.uk/hi/english/static/in_depth/world/2002/disposable_planet/waste/chinese_workshop/.

BBC News. 2003a. "Computers to Africa Scheme Criticized." June 6.

BBC News. 2003b. "Recycling Law Boost Hi-Tech Transfer." June 6.

Beans, Bruce. 2002. "The Waste That Didn't Make Haste." *Washington Post,* July 17.

Beauchamps, Alex. 1998a. "Open Letter to the People of Philadelphia from Haitian Environmentalist." April.

Beauchamps, Alex. 1998b. "Message to RTS List Serve." March 25.

Beauchamps, Alex. 1998c. "Letter to RTS Members." Haiti: COHPEDA, March 24.

Beck, Ulrich. 1992. *Risk Society: Towards a New Modernity.* Thousand Oaks, Calif.: Sage.

Beck, Ulrich. 1995. *Ecological Enlightenment: Essays on the Politics of the Risk Society.* Amherst, N.Y.: Humanity Books.

Beck, Ulrich. 1999. *World Risk Society.* Oxford: Polity Press.

Been, Vicki. 1993a. "Locally Undesirable Land Uses in Minority Neighborhoods." *Yale Law Review* 103:1383.

Been, Vicki. 1993b. "What's Fairness Got to Do with It? Environmental Justice and Siting of Locally Undesirable Land Uses." *Cornell Law Review* 78:1001.

Benton, Ted. 2002. "Social Theory and Ecological Politics: Reflexive Modernization or Green Socialism?" In Riley Dunlap, Frederick Buttel, Peter Dickens, and August Gijswijt (Eds.), *Sociological Theory and the Environment: Classical Foundations, Contemporary Insights.* Lanham, Md.: Rowman and Littlefield.

Bhabha, Homi. 2004. *The Location of Culture.* London: Routledge.

Black, Richard. 2004. "E-Waste Rules Still Being Flouted." *BBC News,* March 19.

Blum, William. 1995. *Killing Hope: U.S. Military and CIA Interventions since World War II.* Monroe, Me.: Common Courage Press.

Blum, William. 2000. *Rogue State: A Guide to the World's Only Super Power.* Monroe, Me.: Common Courage Press.

Boggs, Carl. 2000. *The End of Politics: Corporate Power and the Decline of the Public Sphere.* New York: Guilford Press.

Boli, John, and George M. Thomas. 1999. *World Polity Formation since 1875: World Culture and International Non-Governmental Organizations.* Palo Alto, Calif.: Stanford University Press.

Bonilla-Silva, Eduardo. 2001. *White Supremacy and Racism in the Post–Civil Rights Era.* Boulder, Colo.: Lynne Rienner.

Bonilla-Silva, Eduardo. 2003. *Racism without Racists: Color-Blind Racism and the Persistence of Racial Inequality in the United States.* Lanham, Md.: Rowman and Littlefield.

Boudreau, Vincent. 1996. "Northern Theory, Southern Protest: Opportunity Structure Analysis in Cross-National Perspective." *Mobilization* 2:175–189.

Bray, F. 1994. Agriculture for Developing Nations. *Scientific American,* July: 30–37.

Braynen, Niccole. 1992. Letter to Ann Leonard of Greenpeace International, Embassy of the Commonwealth of the Bahamas, Washington, D.C. March 3.

Brenner, Neil. 2002. "Preface: From the 'New Localism' to the Spaces of Neoliberalism." *Antipode,* 34(3): 341–347.

Brigden, K., I. Labunska, D. Santillo, and M. Allsopp. 2005. *Recycling of Electronic Wastes in China and India: Workplace and Environmental Contamination.* Exeter, UK: Greenpeace Research Laboratories, Department of Biological Sciences, University of Exeter, August.

Brough, David. 2001. "Ethiopia Says Pesticide Dumps Are 'Time Bomb.'" Reuters News Service, April 20.

Brown, Phil, and Faith Ferguson. 1995. "'Making a Big Stink': Women's Work, Women's Relationships, and Toxic Waste Activism." *Gender and Society* 9: 145–172.

Brulle, Robert J., and David N. Pellow. 2006. "Environmental Justice: Human Health and Environmental Inequalities." *Annual Review of Public Health* 27: 103–124.

Bruno, Kenny. 1998a. Letter to Paula Palmer of Global Response, February 3.

Bruno, Kenny. 1998b. "Philly Waste Go Home." *Multinational Monitor,* January–February,.vol 19, no 1 and 2, p. 2.

Bryant, Bunyan (Ed.). 1995. *Environmental Justice: Issues, Policies, and Solutions.* Washington, D.C.: Island Press.

Bryant, Bunyan, and Paul Mohai (Eds.). 1992. *Race and the Incidence of Environmental Hazards: A Time for Discourse.* Boulder, Colo.: Westview Press.

Bullard, Robert. 1992. Keynote address to the Midwest Environmental Justice Conference, Olive Harvey College, Chicago, October.

Bullard, Robert. 2000. *Dumping in Dixie: Race, Class, and Environmental Quality.* Boulder, Colo.: Westview Press.

Bullard, Robert. 2005a. "Environmental Justice in the 21st Century." In Robert Bullard (Ed.), *The Quest for Environmental Justice: Human Rights and the Politics of Pollution.* San Francisco: Sierra Club Books.

Bullard, Robert (Ed.). 2005b. *The Quest for Environmental Justice: Human Rights and the Politics of Pollution.* San Francisco: Sierra Club Books.

Bullard, Robert, Glenn Johnson, and Angel Torres (Eds.). 2000. *Sprawl City: Race, Politics, and Planning in Atlanta.* Washington, D.C.: Island Press.

Bullard, Robert, and Beverly Wright. 1993. "The Effects of Occupational Injury, Illness and Disease on the Health Status of Black Americans." In Richard Hofrichter (Ed.), *Toxic Struggles: The Theory and Practice of Environmental Justice.* Philadelphia: New Society Publisher.

Byster, Leslie. 2001. "Poison PCs: The Growing Environmental Problem." Paper presented at the Waste Not Asia Conference, Thailand, July.

Cahalan, Steve. 2003. "Holmen Firm Pledges to Recycle Waste Electronics." *La Crosse (Wisconsin) Tribune,* February 26.

Cahn, Claude. 2004. "Breakthrough: Challenging Coercive Sterilisations of Romani Women in the Czech Republic." *Roma Rights,* no. 3 and 4:103–112.

Calonzo, Manny. 2004. Invitation from GAIA for GDAW 2004, August 2.

Camacho, David (Ed.). 1998. *Environmental Injustices, Political Struggles: Race, Class, and the Environment.* Durham, N.C.: Duke University Press.

Campbell, Malcolm. 1992. Letter to Lowell Arbury, Abaco Wholesale, March.

Capek, Stella. 1993. "The 'Environmental Justice' Frame: A Conceptual Discussion and an Application." *Social Problems* 40:5–24.

Carson, Rachel. 2002 [1962]. *Silent Spring*. Fortieth Anniversary Edition. New York: Mariner Books.

Centers for Disease Control and Prevention. 2003. *Second National Report on Human Exposure to Environmental Chemicals*. Atlanta, Ga.: Centers for Disease Control and Prevention, January.

Central Intelligence Agency. 2005. *CIA World Factbook: The Bahamas*. November. http://www.cia.gov/cia/publications/factbook/geos/bf.html.

Chang, Iris. 1998. *The Rape of Nanking: The Forgotten Holocaust of World War II*. New York: Penguin.

Chang, Nancy. 2002. *Silencing Political Dissent: How Post September 11 Anti-Terrorism Measures Threaten Our Civil Liberties*. New York: Seven Stories Press.

Charlier, Marj. 1993. "Going South: U.S. Mining Firms, Unwelcome at Home, Flock to Latin America." *Wall Street Journal,* June 18.

"China Accuses U.S. of Dumping Illegal Waste." 1996. *San Francisco Chronicle,* June 1, C1.

Chomsky, Noam. 1993. *Year 501: The Conquest Continues*. Boston: South End Press.

Chua, Amy. 2003. *World on Fire: How Exporting Free Market Democracy Breeds Ethnic Hatred and Global Instability*. New York: Doubleday.

Churchill, Ward, and Jim Vander Wall. 2001. *The COINTELPRO Papers: Documents from the FBI's Secret Wars against Dissent in the United States*. Boston: South End Press.

Clapp, Jennifer. 2001. *Toxic Exports: The Transfer of Hazardous Wastes from Rich to Poor Countries*. Ithaca, N.Y.: Cornell University Press.

Clarke, Tony. 1996. "Mechanisms of Corporate Rule." In J. Mander and E. Goldsmith (Eds.), *The Case against the Global Economy*. San Francisco: Sierra Club Books.

Coalition Press Release. 2000. "Mozambique Activists Win Huge Victory against Toxic Wastey." October 5.http://www.ban.org.

Cohen, Felix S. 1953. "The Erosion of Indian Rights, 1950–1953: A Case Study in Bureaucracy." *Yale Law Journal 62*:348, 390.

Cohen, Gary. 1997. "Eliminating Toxic Chemical Threats—A New Framework." *Global Pesticide Campaigner* 7(4): 16–17.

COHPEDA (Haitian Collective for the Protection of the Environment and an Alternative Development). 1998. Petition to the Parliament and President of the Republic of Haiti, March 29.

"Cold Water Poured on IT's Environmental Pluses." 2003. *Environment Daily,* November 6.

Cole, Luke. 2002. "Expanding Civil Rights Protections in Contested Terrain: Using Title VI of the Civil Rights Act of 1964." In Gary Bryner and Douglass Kenney

(Eds.), *Justice and Natural Resources: Concepts, Strategies, and Applications.* Washington, D.C.: Island Press.

Cole, Luke, and Sheila Foster. 2001. *From the Ground Up: Environmental Racism and the Rise of the Environmental Justice Movement.* New York: New York University Press.

Collins, Patricia Hill. 2005. "Black Public Intellectuals: From DuBois to the Present." *Contexts* 4(4):22–27.

Commoner, Barry. 1971. *The Closing Circle: Nature, Man, and Technology.* New York: Random House.

"Computer Giant HP Mute over Toxin Use." 2005. *China Daily,* July 7.

Computer TakeBack Campaign. 2002. "Dell Shareholders Urged to Examine All Aspects of Company's Performance." Press release, May 1.

Computer TakeBack Campaign. 2003. "The Solutions: Electronics Recyclers Pledge of True Stewardship." June. http://www.computertakeback.com/the_solutions/recycler_s_pledge.cfm.

Connett, Paul. 1998. "Municipal Waste Incineration: A Poor Solution for the Twenty First Century." Presentation at the Fourth Annual International Management Conference on Waste-to-Energy, Amsterdam, November 24–25.

Crenshaw, Kimberle. 2000. "Were the Critics Right about Rights? Reassessing the American Debate about Rights in the Post-Reform Era." In Mahmood Mamdani (Ed.), *Beyond Rights Talk and Culture Talk.* New York: St. Martin's Press.

Critharis, M. 1990. "Third World Nations Are Down in the Dumps: The Exportation of Hazardous Waste." *Brooklyn Journal of International Law* 16:311.

Crowe, David M. 1996. *A History of the Gypsies of Eastern Europe and Russia.* New York: St. Martin's Press.

Darder, Antonia, and Rodolfo D. Torres. 2004. *After Race: Racism after Multiculturalism.* New York: New York University Press.

Dayaneni, Gopal, and John Doucette. 2005. *System Error: Toxic Tech Poisoning People and Planet.* Silicon Valley Toxics Coalition, December. San Jose, California.

Dean, Katie. 2003. "Recyclers Pledge to Cut E-Waste." *Wired News,* February 26.

Delaplane, Keith. 2000. "Pesticide Usage in the United States: History, Benefits, Risks, and Trends." Bulletin 1121, November. Athens: Cooperative Extension Service, University of Georgia College of Agricultural and Environmental Sciences.

"Dell Cancels Contract for Inmate Labor." 2003. *San Diego Union Tribune,* July 5.

Deloria, Vine. 1985. *Behind the Trail of Broken Treaties: An Indian Declaration of Independence.* Austin: University of Texas Press.

Dennis, L. K., and S. L. Pallota. 2001. "Chronic Disease in Rural Health." In Sana Loue and Beth E. Quill (Eds.), *Handbook of Rural Health.* New York: Kluwer.

Department of Government and Foreign Trade. 2001. *Statistics of Foreign Trade of India.* Government of India.

Devraj, Ranjit. 2003. "Indian Coke, Pepsi Laced with Pesticides, Says NGO." Inter Press Service, August 5.

Dewailly E., et al. 1994. "High Organochlorine Body Burden in Women with Estrogen Receptor-Positive Breast Cancer." *Journal of the National Cancer Institute* 86:232–234.

DiMaggio, Paul, and Walter Powell. 1983. "'The Iron Cage Revisited': Institutional Isomorphism and Collective Rationality in Organizational Fields." *American Sociological Review* 48:147–160.

Dinham, Barbara. 1993. *The Pesticide Hazard.* London: Zed Books.

Dinham, Barbara. 1996. "Points on Paraquat." *Pesticides News,* no. 32, June.

Douglass, Frederick. 1881. "The Color Line." *North American Review,* June.

Doyle, Jack. 2004. *Trespass against Us: Dow Chemical and the Toxic Century.* Boston: Common Courage Press and the Environmental Health Fund.

Druley, Laurel. 1997. "Still Dumping after All These Years." *Mother Jones,* November 11.

DuBois, W.E.B. 1977 [1935]. *Black Reconstruction: An Essay Toward a History of the Part Which Black Folk Played in the Attempt to Reconstruct Democracy in America, 1860–1880.* New York: Atheneum.

Duck and Cover. 2003. "Since when did unquestioning obedience to corporate interests become patriotic?" Bumper sticker.

Dunlap, Riley, and Brent K. Marshall. 2006. "Environmental Sociology." In Clifton D. Bryant and Dennis L. Peek (Eds.), *The Handbook of 21st Century Sociology.* Thousand Oaks, Calif.: Sage.

Dussel, Enrique. 1998. " Beyond Eurocentrism: The World-System and the Limits of Modernity." In Fredric Jameson and Masao Miyashi (Eds.), *The Cultures of Globalization.* Durham, N.C.: Duke University Press.

"Eco-Reports: Pesticides in the English Speaking Caribbean." N.d. Insert titled "Obsolete Pesticides in Abaco."

Eddleston, M. 2000. "Patterns and Problems of Deliberate Self-Poisoning in the Developing World." *Quarterly Journal of Medicine* 93:715–731.

Eddleston, Michael, et al. 2002. "Pesticide Poisoning in the Developing World—A Minimum Pesticides List." *Lancet* 360:1163–1167.

Ehrlich, Paul. 1989. "Preface." In Robert van den Bosch (Ed.), *The Pesticide Conspiracy.* Berkeley: University of California Press.

Eisinger, Peter. 1973. "The Conditions of Protest Behavior in American Cities." *American Political Science Review* 67:11–68.

Environment News Service. 1998. [No title.] June 2.

Environmental New Service. 2000. "Mozambique Environmentalists Defeat Incinerator Plan." Maputo, Mozambique, October 13.

Environmental News Service. 2002a. "North American Success Comes at Global Expense." August 13.

Environmental News Service. 2002b. "North America Shifts Pollution from Air to Land." May 31.

Environmental News Service. 2002c. "Overall Toxic Releases Down, Hazwaste Up in 2000." May 28.

Environmental News Service. 2002. "Plans for European Green Electronics Law Roasted." January 22d.

Environmental News Service. 2003a. "UK Household Waste Predicted to Double by 2020." November 29.

Environmental News Service. 2003b. "Phone Switching Rule Will Trash Millions of Cell Phones." November 25.

Environment News Service. 2003c. "Herbicide Paraquat Approved for Sale in Europe." October 6.

Environment News Service. 2003d. "CropLife Pledges $30M to Rid Africa of Obsolete Pesticides." Washington, D.C., August 13.

Environmental News Service. 2003e. "Toxic Substances Put One in Five EU Workers at Risk." May 16.

Environment News Service. 2003f. "Philippines: People's Congress Urges Land, Food without Poisons." April 7.

Environmental News Service. 2003g. "Britain Confronts Impending Hazwaste Crisis." April 2.

Environment News Service. 2004a. "Rich Nations Gobbling Resources at an Unsustainable Rate." March 30.

Environment News Service. 2004b. "One Billion Computers Devouring Earth's Resources." March 8.

Environment News Service. 2004c. "Mercury Levels Too High in One in Six Pregnant Women." February 9.

Environment News Service. 2005. "Rubbish Illegally Dumped in England Every 35 Seconds." March 2.

Epstein, Samuel S. 1979. *The Politics of Cancer.* New York: Anchor Books.

Espiritu, Yen Le. 2003. *Homebound: Filipino American Lives across Cultures, Communities, and Countries.* Berkeley: University of California Press.

Essential Action. 1998. March. "Protests in Haiti over Ash Continue." *Essential Action News.* Washington, D.C.

Essential Action. 1999. "University Students Tell Rendell to Get His Ash Home!" Essential Action Press release.

Estrada, Daniela. 2004. "Electronic Garbage Poses Increasing Dangers." Inter-Press Service News Agency (Tierramérica), Santiago, Chile, November 9.

European Commission's Scientific Committee of Plants. 2002. Minutes of the 31st meeting of the Scientific Committee of Plants, Brussels, Belgium, December 20.

Faber, Daniel (Ed.). 1998. *The Struggle for Ecological Democracy: Environmental Justice Movements in the United States*. New York: Guilford Press.

Faber, Daniel, and Eric Krieg. 2001. *Unequal Exposure to Ecological Hazards: Environmental Injustices in the Commonwealth of Massachusetts*. Boston: Northeastern University.

Faison, Seth. 1997. "China Convicts American as Trash Smuggler." *New York Times,* January 14, A3.

Fawzi, Mohamed, and Izzat Abdul Razaq. 2004. "US. Dump in Baghdad: Some See Treasure, Others See Trash." Environment News Service, January 20.

Feagin, Joe, and Hernan Vera 2001. *Liberation Sociology.* Boulder, Colo.: Westview Press.

Feagin, Joe, Hernan Vera, and Pinar Batur. 2001. *White Racism.* 2nd ed. New York: Routledge.

Fenton Communications. 1999. "National Ad Campaign Charges Texaco with Race Discrimination—This time in Amazon Rainforest." September 23. http://www.texacorainforest.com/press/sept23_99

Fernandez, Irene. 2002. "Pesticide Action Network Asia Pacific (PANAP) Study." Malaysia, March 21.

Fiorillo, Victor, and Liz Spikol. 2001. "Ashes to Ashes, Dust to Dust." *Philadelphia Weekly,* January.

Fischer, Frank. 2000. *Citizens, Experts, and the Environment: The Politics of Local Knowledge.* Durham, N.C.: Duke University Press.

Fisher, Dana and W. Freudenburg. 2001. "Ecological Modernization and Its Critics." *Society and Natural Resources* 14:701–709.

Florini, Ann M. (Ed.). 2000. *The Third Force: The Rise of Transnational Civil Society.* Washington, D.C.: Carnegie Endowment for International Peace.

Flynn, Laurie. 2004. "2 PC Makers Favor Bigger Recycling Roles." *New York Times,* May 19.

Foster, John Bellamy. 2000. *Marx's Ecology: Materialism and Nature.* New York: Monthly Review Press.

Foster, John Bellamy. 2006. "Aspects of Class in the United States: An Introduction." *Monthly Review* 58(3): 6–11.

Fox, Jonathan A., and L. D. Brown. 1998. *The Struggle for Accountability: The World Bank, NGOs, and Grassroots Movements.* Cambridge, Mass.: The MIT Press.

Fox, Steve. 1992. *Toxic Work: Women Workers at GTE Lenkurt.* Philadelphia: Temple University Press.

Freeman, Aaron. 1996. "Sanctioning Burma." *Multinational Monitor* 17 (6): 2.

Frey, R. Scott. 1994. "The International Traffic in Hazardous Wastes." *Journal of Environmental Systems* 23:165–177.

Frey, Scott R. 1995. "The International Traffic in Pesticides." *Technological Forecasting and Social Change* 50:151–169.

Frey, R. Scott. 1998. "The Export of Hazardous Industries to the Peripheral Zones of the World-System." *Journal of Developing Societies* 41:66–81.

Frey, R. Scott. 2001. "The Hazardous Waste Stream in the World-System" In R. Scott Frey (Ed.), *The Environment and Society Reader.* Boston: Allyn and Bacon.

Friedman, Thomas. 2005. *The World Is Flat: A Brief History of the Twenty-First Century.* New York: Farrar, Straus, and Giroux.

Friends of the Earth UK. 2003. "Waste: The Mess We're In!" http://www.foe.co.uk.

GAIA. 2001. "Mixed, Stinking Garbage of Metro Officials Exposed." Manila, October 26.

Garcia, Ana. 2003. "Pesticide Exposure and Women's Health." *American Journal of Industrial Medicine.* 44(2):584–594.

Garcia, Connie, and 2006. "Community-Based Organizing for Labor Rights, Health and the Environment: Television Manufacturing on the Mexico-U.S. Border." In Ted Smith, David Sonnenfeld, and David N. Pellow (Eds.), *Challenging the Chip: Labor Rights and Environmental Justice in the Global Electronics Industry.* Philadelphia: Temple University Press.

Garcia-Johnson, Ronie. 2002. *Exporting Environmentalism: U.S. Multinational Chemical Corporations in Brazil and Mexico.* Cambridge, Mass.: The MIT Press.

Gbadegesinn, Segun. 2001. "Multinational Corporations, Developed Nations, and Environmental Racism: Toxic Waste, Oil Exploration, and Eco-Catastrophe." In Laura Westra and Bill E. Lawson (Eds.), *Faces of Environmental Racism: Confronting Issues of Global Justice.* 2nd ed. Lanham, Md.: Rowman and Littlefield.

Gedicks, Al. 1993. *The New Resource Wars: Native and Environmental Struggles against Multinational Corporations.* Boston: South End Press.

Gedicks, Al. 2001. *Resource Rebels: Native Challenges to Mining and Oil Corporations.* Boston: South End Press.

Getches, David, and David Pellow. 2002. "Beyond 'Traditional' Environmental Justice: How Large a Tent?" In Kathryn Mutz, Gary Bryner, and Douglas Kenney (Eds.), *Justice and Natural Resources: Concepts, Strategies, and Applications.* Washington, D.C.: Island Press.

Ghosh, Aditya. 2006. "City Is Now Biggest Electronic Graveyard." DNA India.com, January 1.

Gilroy, Paul. 2000. *Against Race: Imagining Political Culture beyond the Color Line.* Cambridge, Mass.: Harvard University Press.

Gilroy, Paul. 2001. *The Black Atlantic: Modernity and Double-Consciousness.* Cambridge, Mass.: Harvard University Press.

Gladstone, Thurston. 1991. "Toxin Leaking into Water Supply, MP Claims." *Nassau Tribune,* March 25.

Gladstone, Thurston. 1992a. "US Company to Remove Toxic Pesticides from Abaco." *Nassau Guardian,* August 10.

Gladstone, Thurston. 1992b. "Abaconians Threaten to Ship Deadly Dioxin to NP." *Nassau Guardian,* April 30.

Gladstone, Thurston. 1992c. "Toxic Substance Threatens Abaco Water Table." *Nassau Guardian,* February 11.

Global Pesticide Campaigner. 2002. "Funds for Clearing Toxic Pesticide Stockpiles in Africa." volume 12, no. 3, December.

Global Pesticide Campaigner. 2003. "War and Pesticides." Volume 13, Number 1 http://www.panna.org/resources/gpc/gpc_200304.13.1.dv.html.

Global Response. 1992a. "Young Environmentalist's Action." June.

Global Response. 1992b. Action #6/92: Toxic Exports—Pesticides / U.S.-Bahamas. June.

Global Response. 1994. *GR Action #7/94,* Pepsico/Plastic-Pollution, July.

Global Response. 2005. GR Action #4/05. April. www.globalresponse.org.

Global Response. 2006. "News Flash. Global Response Victory! Roma Families Moved to Safety." August.

Global Response. 1998. Action #2/98, Project 'Return to Sender'. February.

Glueck, Grace. 2003. "Subject Is U.S., Object Is Art." *New York Time,* July 4, B25.

Goldberg, David Theo. 2002. *The Racial State.* Malden, Mass.: Blackwell.

Goldfrank, Walter, David Goodman, and Andrew Szasz (Eds.). 1999. *Ecology and the World System.* Westport, Conn.: Greenwood Press.

Goldman, Benjamin. 1991. *The Truth about Where You Live: An Atlas for Action on Toxins and Mortality.* New York: Times Books/Random House.

Goldman, Michael (Ed.). 1998. *Privatizing Nature: Political Struggles for the Global Commons.* New Brunswick, N.J.: Rutgers University Press.

Goldsmith, Marsha F. 1989. "As Farmworkers Help Keep America Healthy, Illness May Be Their Harvest." *Journal of American Medical Association* 261(22): 3201–3213.

Goldstein, N., and Madtes, C. 2001. "The State of Garbage in America." *Biocycle Magazine,* December, 42–54.

Goodman, Peter S. 2003. "China Serves as Dump Site for Computers." *Washington Post Foreign Service,* February 24.

Gordon, Holly, and Harley, Keith. 2005. "Environmental Justice and the Legal System." In David Naguib Pellow and Robert J. Brulle (Eds.), *Power, Justice, and the Environment: A Critical Appraisal of the Environmental Justice Movement.* Cambridge, Mass.: The MIT Press.

Gottlieb, Robert. 1993. *Forcing the Spring: The Transformation of the American Environmental Movement.* Washington, D.C.: Island Press.

Gottlieb, Robert. 2001. *Environmentalism Unbound: Exploring New Pathways for Change.* Cambridge, Mass.: MIT Press.

Gough, Neil. 2002. "Garbage In, Garbage Out: Castoffs from the Computer Age Are a Financial Windfall for Chinese Villagers. But at What Cost?" *Time Asia,* March 11.

Gould, Kenneth. 2003. "Transnational Environmentalism, Power and Development in Belize" *Belizean Studies.* 25(2): 59–70.

Gould, Kenneth Alan. 2005. "Comments on Environmental Justice Movements." Panel on Environmental Justice Movements. Sociologists Without Borders Conference, Philadelphia, August.

Gould, Kenneth, Allan Schnaiberg, and Adam Weinberg. 1996. *Local Environmental Struggles: Citizen Activism in the Treadmill of Production.* Cambridge: Cambridge University Press.

GrassRoots Recycling Network. 2003a. "EPA: Keep Toxic PC's out of Asia." Madison, Wis.: GrassRoots Recycling Network, July 6.

GrassRoots Recycling Network. 2003b. "GRRN Mission." July 6. http://www.grrn.org.

Greenpeace. 1992. *We've Been Had! Seattle's Plastic Recycling Myth.* Seattle, Wash.: Greenpeace International Waste Trade Project, June.

Greenpeace. 1994. "Pepsi Action Alert: Pepsi Pollutes the Planet with Plastic." August.

Greenpeace. 1998a. *Toxic Legacies; Poisoned Futures.* November.

Greenpeace. 1998b. "Basel Ban: Final Round to Put an End to Global Waste Dumping." Press release, February 23.

Greenpeace. 1998c. "Dutch PVC Waste Still Exported to Asia Despite International Agreement, Greenpeace Calls for an End to 'Delayed Dumping.'" Press release, February 4. http://www.greenpeace.org/pressreleases/toxics/1998feb4.html.

Greenpeace. 1998d. "On Anniversary of an Environmental Scandal, Plans Revealed to Bring Philadelphia's Toxic Waste back from Haiti." Press release, January 14.

Greenpeace. 1999a. "First Ever National Incinerator Ban Signed Today in the Philippines." Press release, June 23. http://www.greenpeace.org/pressreleases/toxics/1999june23.html.

Greenpeace. 1999b. "Solvay Chemical Plant in Brazil Identified as Source of European Food Contamination." March 25.

Greenpeace. 2000. "Japan Uses Aid to Export 'Toxic' Incineration." Press release, February 8.

Greenpeace. 2001. "Task Force Forming to Oppose Cherokee Nation Trash Deal." Press release, September 25.

"Greenpeace Indicts Pepsi for Dumping Waste in India." 1994. *Business Standard,* October 20.

Greenpeace International. 1998. "Danish Development Project Encourages Toxic Waste Trade into Mozambique." Press release, October 6.

"Greenpeace Launches Campaign against E-Waste." 2005. September 6. http://www.newindpress.com.

Greenpeace USA. 2006a. "Hewlett Packard Changes Its Polluting Practices." March 15. www.greenpeaceusa.org.

Greenpeace USA. 2006b. Membership solicitation letter, Washington, D.C., January.

Greenpeace USA. N.d. "Disposal of Obsolete Pesticide Stocks in West Africa." Press release, Washington, D.C.

Greenpeace and Basel Action Network. N.d. "Key Findings from Taizhou Field Investigation."

Grossman, Elizabeth. 2006. *High Tech Trash: Digital Devices, Hidden Toxics, and Human Health.* Washington, D.C.: Island Press.

Guidry, John A., Michael D. Kennedy, and Mayer N. Zald (Eds.). 2000. *Globalizations and Social Movements: Culture, Power, and the Transnational Public Sphere.* Ann Arbor: University of Michigan Press.

Guinier, Lani, and Gerald Torres. 2002. *The Miner's Canary: Enlisting Race, Resisting Power, Transforming Democracy.* Cambridge, Mass.: Harvard University Press.

Hacker, Andrew (Ed.). 1965. *The Corporation Take-Over.* New York: Doubleday.

Hall, Stuart. 1978. "The Politics of Mugging." In Stuart Hall (Ed.), *Policing the Crisis: Mugging, the State, and Law and Order.* New York: Holmes and Meier.

Hanlon, Joseph. 2001. "Bank Corruption Becomes Site of Struggle in Mozambique." *Review of African Political Economy* 91: 53–72.

Hannaford, Ian. 1996. *Race: The History of an Idea in the West.* Baltimore: Johns Hopkins University Press.

Harden, Monique. 2005. Presentation at the Transatlantic Initiative to Promote Environmental Justice, Central European University, Budapest, Hungary, October 30.

Hawley, Susan. 2005. "Redesigning the Northern State to Combat Global Corruption." Plenary presentation at The Cornerhouse: "Redesigning the State: Political Corruption in Development Policy and Practice," University of Manchester, November 25.

"HC Stays Order on Pepsico License Cancellation." 2004. Domain-b.com. April 24. http://www.domain-b.com/companies/companies_p/pepsiindia/20040424_licence.html.

Heath Care Without Harm. N.D. *How to Shut Down an Incinerator.* Washington, D.C.

Hernandez, Von. 2002. Presentation at the Global Symposium on the High-Tech Industry, San Jose, Calif., November.

Higgins, Robert. 1994. "Race, Pollution and the Mastery of Nature." *Environmental Ethics* 16:251–263.

Hilz, Christopher. 1992. *The International Toxic Waste Trade.* New York: Van Nostrand Reinhold.

Hochshild, Adam. 2005. *Bury the Chains: Prophets and Rebels in the Fight to Free and Empire's Slaves.* Boston: Houghton Mifflin.

Hoff, John. 1992. Letter to Global Response, July 24.

Hooks, Gregory, and Chad Smith. 2004. "The Treadmill of Destruction: National Sacrifice Areas and Native Americans." *American Sociological Review* 69:558–575.

Horrigan, Alice, and Jim Motavalli. 1997. "Talking Trash." *E Magazine* 8(2). http://www.emagazine.com/view/?409&src=QHA241.

Horsman, Reginald. 1981. *Race and Manifest Destiny: The Origins of American Racial Anglo-Saxonism.* Cambridge, Mass.: Harvard University Press.

Hossfeld, Karen. 1990. "'Their Logic against Them': Contradictions in Sex, Race, and Class in Silicon Valley." In Kathryn Ward (Ed.), *Women Workers and Global Restructuring.* Ithaca, N.Y.: Institute for Labor Relations Press.

Hough, Peter. 2000. "Return to Sender: Regulating the International Trade in Pesticides." Paper presented at the ECPR Workshop on The Politics of Food, Copenhagen, Denmark, April 14–19.

Hua, Julietta. 2006. *Women's Rights as Human Rights? Problematizing the 'Third World Woman' as Global Subject.* Unpublished dissertation thesis, University of California, San Diego.

Huber, J. 1982. *Die verlorene Unschuld der Okologie: Neue Technologien und Superindustrielle Entwicklung.* Frankfurt am Main: Fischer.

Human Rights Watch. 1999. China and Tibet. *World Report.* Geneva.

Hurley, Andrew. 1995. *Environmental Inequalities: Class, Race, and Industrial Pollution in Gary, Indiana, 1945–1980.* Chapel Hill: University of North Carolina Press.

Iles, Alastair. 2004. "Mapping Environmental Justice in Technology Flows: Computer Waste Impacts in Asia." *Global Environmental Politics* 4(4):76–107.

International Labor Office. 1994. "Chemicals in the Working environment." In *World Labor Report.* Geneva: International Labor Organization.

International Labor Organization. 1996. "Wage Workers in Agriculture: Conditions of Employment and Work." Sectoral activities program, Geneva.

Isin, E. H. 1998. "Governing Toronto without Government: Liberalism and Neoliberalism." *Studies in Political Economy* 56:169–191.

Jaitly, Jaya, and Kavalijit Singh. 1994. "PepsiCo and the India Government's Betrayal of the People of India." Press release, November 9.

James, Joy. 1996. *Resisting State Violence: Radicalism, Gender, and Race in U.S. Culture.* Minneapolis: University of Minnesota Press.

"Japan: Philippines Case Bares Inadequacy of Waste Rules." 2000. *Yomiuri Shimbun,* January 12.

Japan Economic Newswire Plus. 1994. "Greenpeace Blocks Ship Carrying Toxic Waste to Manila." March 2.

Jeyaratnam, J. 1990. "Acute Pesticide Poisoning: A Major Global Health Problem." *World Health Statistics Quarterly* 43:139–144.

Jeyaratnam, J., K. C. Lun, and W. O. Phoon. 1987. "Survey of Acute Pesticide Poisoning among Agricultural Workers in Four Asian Countries." *Bulletin of the World Health Organization* 65:521–527

Johnson, Chalmers. 2000. *Blowback: The Costs and Consequences of American Empire.* New York: Holt.

Johnson, Chalmers. 2003. "The Looting of Asia." *London Review of Books,* November 20, vol. 25, no. 22.

Johnson, Chalmers. 2004. *The Sorrows of Empire: Militarism, Secrecy, and the End of the Republic.* New York: Metropolitan Books.

Kamdar, Seema. 2004. "E-Waste Gives Citizens a Big Headache." *India Times News Network,* July 19.

Kanyegirire, Andrew. 2003. "The Throw-Away Factor." *New Vision.* January 16.

Karshan, Michelle. 2000. "The Haitian People Achieve Environmental Justice for Earth Day." Press release, Haitian government, April 22.

Kauffman, Daniel. 2004. "Corruption, Governance and Security: The Challenges for Rich Countries and the World." In World Bank, *Global Competitiveness Report.* December. Geneva, Switzerland: World Economic Forum.

Keck, Margaret E., and Sikkink, Kathryn. 1998. *Activists beyond Borders: Advocacy Networks in International Politics.* Ithaca, N.Y.: Cornell University Press.

Kennedy, Robert F. Jr. 2003. "Crimes against Nature." *Rolling Stone,* December 11

Khagram, Sanjeev, James Riker, and Kathryn Sikkink. 2002. "From Santiago to Seattle." In Sanjeev Khagram, James V. Riker, and Kathryn Sikkink (Eds.), *Restructuring World Politics: Transnational Social Movements, Networks, and Norms.* Minneapolis: University of Minnesota Press.

Kirkpatrick, Sale. 1990. *Conquest of Paradise.* New York: Knopf.

Kitschelt, Herbert. 1986. "Political Opportunity Structures and Political Protest: Anti-Nuclear Movements in Four Democracies." *British Journal of Political Science* 16:57–85.

Kivel, Paul. 2002. *Uprooting Racism: How White People Can Work for Racial Justice.* Philadelphia: New Society Publishers.

Klein, Naomi. 2005. "A Noose Not a Bracelet: Africa Is a Rich Continent Made Poor by Rapacious Western Corporations." *Guardian,* June 10.

Knight, Danielle. 2000. "Sony Monitoring Environmental Activists." InterPress Service, September 22.

Knight, Danielle. 2001. "Controversy around Mercury Shipment from US to India." InterPress Service, January 25.

Konefal, Jason, and Michael Mascarenhas. 2005. "The Shifting Political Economy of the Global Agrifood System: Consumption and the Treadmill of Production." *Berkeley Journal of Sociology* 49: 76–96.

Kovel, Joel. 1984. *White Racism: A Psychohistory.* New York: Columbia University Press.

Krieg, E. 1998. "The Two Faces of Toxic Waste." *Sociological Forum* 13:3–20.

Kriesberg, Louis. 1997. "Social Movements and Global Transformation." In J. Smith, C. Chatfield, and R. Pagnucco (Eds.), *Transnational Social Movements and Global Politics: Solidarity Beyond the State.* Syracuse, N.Y.: Syracuse University Press.

Krishnamoorthy, Bala. 2003. "Waste Management Practices in the Hospitals in Mumbai." *Vishitri: The Journal of the Jamnalal Bajaj Institute of Management Studies,* April. http:www.jbims.edu/vishitri_2.htm.

Kucinskaite, Egle Kristina. 2002. "In the Dark: Segregating Roma in Lithuania." *Roma Rights,* no. 3 and 4:26–29.

Kuhndt, M., J. von Geibler, V. Turk, S. Moll, K. O. Schallabock, and S. Steger. 2003. *Virtual Dematerialisation. E-Business and Factor X.* Digital Europe and the Wupperthal Institute, March. Berlin, Germany.

LaDuke, Winona. 1999. *All Our Relations: Native Struggles for Land and Life.* Cambridge, Mass.: South End Press.

Larsen, Suzie. 1998. "In the Clinton Era, U.S. Dumping of Hazardous Pesticides Overseas Has Gotten Worse." *Nation,* July 21.

Lavelle, Marianne, and Marcia Coyle. 1992. "Unequal Protection: The Racial Divide in Environmental Law." *National Law Journal* 15:S1–S12.

Lavery, Brian. 2005. "Ireland's Garbage Secrets Come Glaringly to Light." *New York Times,* July 10.

Lee, Jennifer. 2004. "Study Clears Pesticide Tests with Humans." *New York Times,* February 20.

Lemos, Anabela. 2003. Mozambique report. GAIA Global Meeting 2003, Penang, Malaysia, March 17–21.

Lemos, Anabela, and Nityanand Jayaraman. 2002. "Killing with the Cure: World Bank's Role in Africa Pesticides Cleanup Raises New Threats." *Economic Justice News Online* 4(4). http://www.50years.org/cms/ejn/v4n4.

Leo, Peter. 2006. "Cell Phone Statistics That May Surprise You." *Pittsburgh Post-Gazette,* March 16.

Leonard, Ann. 1991. Letter to Joseph Curry, February 2, Greenpeace International.

Leonard, Ann. 1992. Letter to Edward St. George, April 14, Greenpeace International Waste Trade Project.

Leonard, Ann. 1993. "South Asia: The New Target of International Waste Traders." *Multinational Monitor,* December. 12–16.

Leonard, Ann. 1994. "Dumping Pepsi's Plastic." *Multinational Monitor,* September 7–10.

Leonard, Ann. 1998. Letter to RTS listserv, March 9.

Leonard, Ann. 1999. Letter to RTS listserv, June 22.

Leonard, Ann. 2005a. "The Global Anti-Incinerator Alliance." Presentation at the Transatlantic Initiative to Promote Environmental Justice, Central European University, Budapest, Hungary, October 29.

Leonard, Ann. 2005b. "GAIA Case Example." Transatlantic Initiative on Environmental Justice, Central European University, Budapest, Hungary, October 28. http://calcultures.ucsd.edu/transatlantic_initiative/papers.htm.

Leong, Apo, and Sanjiv Pandita. 2006. "'Made in China': Electronics Workers in the World's Fastest Growing Economy." In Ted Smith, David Sonnenfeld, and David N. Pellow (Eds.), *Challenging the Chip: Labor Rights and Environmental Justice in the Global Electronics Industry.* Philadelphia: Temple University Press.

Lerner, Steve. 2005. *Diamond: A Struggle for Environmental Justice in Louisiana's Chemical Corridor.* Cambridge, Mass.: The MIT Press.

Levenstein, Charles, and John Wooding. 1998. "Dying for a Living: Workers, Production, and the Environment." In Daniel Faber (Ed.), *The Struggle for Ecological Democracy: Environmental Justice Movements in the United States.* New York: Guilford Press.

Liden, Jon. 1995. "In Malaysia, Association Fights Dumping, Expropriations, 'Bad Development.'" *International Herald Tribune.* February 20.

Lipsitz, George. 2001. *American Studies in a Moment of Danger.* Minneapolis: University of Minnesota Press.

Lipsitz, George. 2005. *The Possessive Investment in Whiteness.* 2nd ed. Philadelphia: Temple University Press.

Liptak, Adam. 2004. "Federal Judge Rules Chemicals Used in Executions Are Humane." *New York Times,* February 7.

London, Leslie, Sylvie de Grosbois, Catharina Wesseling, Sophia Kisting, Hanna Andrea Rother, and Donna Mergler. 2002. "Pesticide Usage and Health Consequences for Women in Developing Countries: Out of Sight, Out of Mind?" *International Journal of Occupational and Environmental Health* 8 (1):45–59.

Lorde, Audre. 1984. *Sister Outsider.* Trumansburg, N.Y.: Crossing Press.

Lowe, Lisa. 1996. *Immigrant Acts: On Asian American Cultural Politics.* Durham, N.C.: Duke University Press.

Lowe, Sarah. 2003. "Toxic Waste Victory in Mozambique." *Horizons* (Oxfam News Magazine on-line), February. http://www.oxfam.org.au/oxfamnews/february_2003/mozambique.html.

Lunder, Sonya, and Renee Sharp. N.d. "Toxic Fire Retardants Building Up in San Francisco Bay Fish." Environmental Working Group. http://www.ewg.org/reports/taintedcatch/es.php.

Madeley, John. 2002. "Paraquat—Syngenta's Controversial Herbicide." Report for Berne Declaration and others. April.

Magubane, Bernard. 1990. *The Political Economy of Race and Class in South America.* New York: Monthly Review Press.

Majumder, Sanjoy. 2006. "Indian state bans Pepsi and Coke." BBC News, Delhi. August 9.

Mangwiro, Charles. 1999. "Obsolete Pesticides Leave Mozambicans with $600,000 Problem." African Eye News Service, South Africa, July 22.

Mander, Jerry and Edward Goldsmith (Eds.). 2001. *The Case against the Global Economy: And for a Turn towards Localization.* London: Kogan Page.

Manheim, Jarol. 2001. *The Death of a Thousand Cuts: Corporate Campaigns and the Attack on the Corporation.* Mahwah, N.J.: Erlbaum.

Marable, Manning. 1983. *How Capitalism Underdeveloped Black America.* Boulder, Colo.: Westview Press.

Marable, Manning. 2004. "Globalization and Racialization." ZNet, August 13. http://www.zmag.org/content/print_article.cfm?itemID=6034§ionID=30

Marbury, Hugh. 1995. "Hazardous Waste Exportation: The Global Manifestation of Environmental Racism." *Vanderbilt Journal of Transnational Law* 28: 251.

Marks, Gary, and Doug McAdam. 1996. "Social Movements and the Changing Structure of Political Opportunity in the European Union." *West European Politics* 19:249–278.

Marx, Karl. 1962. *Theses on Feuerbach.* Reprinted in Karl Marx and Frederick Engels, *Selected Works,* vol. 2 . Moscow: Foreign Languages Publishing House.

Massey, Douglas, and Nancy Denton. 1993. *American Apartheid: Segregation and the Making of the Underclass.* Cambridge, Mass.: Harvard University Press.

May, Kevin. 2006. Statement of Kevin May, Greenpeace China, to Hewlett-Packard Annual Shareholders Meeting, Los Angeles, March 15.

McAdam, Doug. 1996. "The Framing Function of Movement Tactics: Strategic Dramaturgy in the American Civil Rights Movement." In Doug McAdam, John McCarthy, and Mayer Zald (Eds.), *Comparative Perspectives on Social Movements.* Cambridge: Cambridge University Press.

McAdam, Doug. 1999. *Political Process and the Development of Black Insurgency, 1930–1970.* Chicago: University of Chicago Press.

McAdam, Doug, Sidney Tarrow, and Charles Tilly (Eds.). 2001. *Dynamics of Contention.* Cambridge: Cambridge University Press.

McCarthy, John. 1997. "The Globalization of Social Movement Theory." In Jackie Smith, Charles Chatfield, and Ron Pagnucco (Eds.), *Transnational Social*

Movements and Global Politics: Solidarity Beyond the State. Syracuse, N.Y.: Syracuse University Press.

McCarthy, John D., and Mayer Zald. 1977. "Resource Mobilization and Social Movements: A Partial Theory." *American Journal of Sociology* 82:1212–1241.

McClintock, Anne. 1995. *Imperial Leather: Race, Gender and Sexuality in the Colonial Conquest.* New York: Routledge.

McDermott, John. 1991. *Corporate Society: Class, Property, and Contemporary Capitalism.* Boulder, Colo.: Westview Press.

McDonald, David (Ed.). 2002. *Environmental Justice in South Africa.* Athens: Ohio University Press.

Mediafax. 2002. "Livaningo Defends Rigorous Study." September 27.

Melosi, Martin. 2001. *Effluent America: Cities, Industry, Energy and the Environment.* Pittsburgh: University of Pittsburgh Press.

Merchant, Carolyn. 1980. *The Death of Nature: Women, Ecology and the Scientific Revolution.* San Francisco: Harper.

Mills, Charles W. 1997. *The Racial Contract.* Ithaca, N.Y.: Cornell University Press.

Mills, Charles W. 2001. "Black Trash." Laura Westra and Bill E. Lawson (Eds.), 2001. *Faces of Environmental Racism: Confronting Issues of Global Justice.* 2nd ed. Lanham, Md.: Rowman and Littlefield.

Ministry of Health. 1992. Untitled document. The Bahamas. Directors Offices, January 19, 27.

Mohanty, Chandra. 2003. *Feminism without Borders: Decolonizing Theory, Practicing Solidarity.* Durham, N.C.: Duke University Press.

Mokhiber, Russell. 1994. "Ten Worst Corporations of 1994." *Multinational Monitor,* December.

Mol, Arthur P. J. 1995. *The Refinement of Production: Ecological Modernization Theory and the Dutch Chemical Industry.* Ultrecht: Jan van Arkel/International Books.

Mol, Arthur. 1996. "Ecological Modernization and Institutional Reflexivity: Environmental Reform in the Late Modern Age." *Environmental Politics* 5(2):302–323.

Mol, Arthur. 2003. *Globalization and Environmental Reform: The Ecological Modernization of the Global Economy.* Cambridge, Mass.: The MIT Press.

Mol, Arthur P. J., and David A. Sonnenfeld. (Eds.). 2000. *Ecological Modernization Around the World.* London: Frank Cass.

Montague, Peter. 1987. "Garbage Ship Wanders the Oceans Searching for Safe Place to Dump." *Rachel's Environment and Health Weekly,* no. 55, December 14.

Montague, Peter. 1998a. "Philadelphia Dumps on the Poor." *Rachel's Environment and Health Weekly,* no. 595, April 23.

Montague, Peter. 1998b. "Incineration News." *Rachel's Environment and Health Weekly,* no. 592, April 2, 4.

Montague, Peter. 2003. "The Chemical Wars." *New Solutions* 14(1):19–41.

Moore, Donald, Jake Kosek, and Anand Pandian (Eds.). 2003. *Race, Nature and the Politics of Difference.* Durham, N.C.: Duke University Press.

Moore, Molly. 2005. "As Youth Riots Spread across France, Muslim Groups Attempt to Intervene." *Washington Post Foreign Service,* November 4.

Morel, Daniel. 1999. "Town Protests Tons of U.S. Toxic Waste Dumped in 1986." Associated Press, Gonaives, Haiti.

Morello-Frosch, Rachel, Manuel Pastor, Jr., Carlos Porras, and James Sadd. 2002. "Environmental Justice and Regional Inequality in Southern California: Implications for Future Research." *Environmental Health Perspectives* 110 (suppl. 2):149–154.

"More Were Exposed to Agent Orange." 2004. *New York Times,* April 17.

Morris, Aldon. 1984. *Origins of the Civil Rights Movement.* New York: Free Press.

Mosley, Walter. 1992. *White Butterfly.* New York: Washington Square Press.

Moses, Marion. 1993. "Farmworkers and Pesticides." In Robert Bullard (Ed.), *Confronting Environmental Racism: Voices from the Grassroots.* Boulder, Colo.: Westview Press.

Moyers, Bill. 1990. *Global Dumping Ground: The International Traffic in Hazardous Wastes.* Santa Ana, California: Seven Locks Press and the Center for Investigative Reporting.

Mpanya, Mutombo. 1992. "The Dumping of Toxic Waste in African Countries: A Case of Poverty and Racism." In Bunyan Bryant and Paul Mohai (Eds.), *Race and the Incidence of Environmental Hazards.* Boulder, Colo.: Westview Press.

Multinationals Resource Center and Health Care Without Harm. 1999. *The World Bank's Dangerous Medicine: Promoting Medical Waste Incineration in Third World Countries.* Washington, D.C., June

"More Were Exposed to Agent Orange." 2004. *New York Time,* April 17.

Nace, Ted. 2005. *Gangs of America: The Rise of Corporate Power and the Disabling of Democracy.* San Francisco: Berrett-Koehler.

National Asian Pacific American Legal Consortium. 2001. *Backlash: Final Report: 2001 Audit of Violence against Asian Americans.* National Asian Pacific American Legal Consortium. Washington, D.C.

Neilsen, Poul. 1999. Letter to Livaningo. January.

Newell, Peter. 2005. "Race, Class, and the Global Politics of Environmental Inequality." *Global Environmental Politics* 5(3): 70–94.

O'Connor, James. 1991. "On the Two Contradictions of Capitalism." *Capitalism, Nature, Socialism* 2(3):107–109.

O'Neill, Kate. 2000. *Waste Trading Among Rich Nations.* Cambridge, Mass.: The MIT Press.

O'Rourke, Dara. 2005. "Market Movements: Nongovernmental Organization Strategies to Influence Global Production and Consumption." *Journal of Industrial Ecology* 9 (1-1): 115–128.

Okihiro, Gary. 1994. *Margins and Mainstreams: Asians in American History and Culture*. Seattle: University of Washington Press.

Omi, Michael, and Howard Winant. 1994. *Racial Formation in the United States: From the 1960s to the 1990s*. 2nd ed. New York: Routledge.

Ong, Aihwa. 2004. "The Chinese Axis: Zoning Technologies and Variegated Sovereignty." *Journal of East Asian Studies* 4:69–96.

Organization for Economic Cooperation and Development. 2001. *Environmental Outlook*. Brussels: OECD.

Osborn, Andrew. 2002. "Britain Accepts Recycling Deal." *Guardian*, October 12.

Palmer, Paula. 1998. Letter to Global Response's Quick Response Network, April 30.

Palmer, Paula. 2005. "The Pen Is Mightier Than the Sword: Global Environmental Justice One Letter at a Time." In David N. Pellow and Robert J. Brulle (Eds.), *Power, Justice, and the Environment: A Critical Appraisal of the Environmental Justice Movement*. Cambridge, Mass.: The MIT Press.

Pandita, Sanjiv. 2006. "Electronics Workers in India." In Ted Smith, David Sonnenfeld, and David N. Pellow (Eds.), *Challenging the Chip: Labor Rights and Environmental Justice in the Global Electronics Industry*. Philadelphia: Temple University Press.

Park, Robert. 1924. "The Concept of Social Distance." *Journal of Applied Sociology* 8:340.

People against Foreign NGO Neocolonialism. 2003. "Unheard Rainforest Conservation Voices from Papua New Guinea." April 7. www.forests.org.

Pegg, J. R. 2004. "State of the World Report: Rising Consumption Unsustainable." Environmental News Service, January 8.

Pellow, David N. 2001. "Environmental Justice and the Political Process: Movements, Corporations, and the State." *Sociological Quarterly* 42:47–67.

Pellow, David N. 2002. *Garbage Wars: The Struggle for Environmental Justice in Chicago*. Cambridge, Mass.: The MIT Press.

Pellow, David Naguib, and Robert J. Brulle (Eds.). *Power, Justice, and the Environment: A Critical Appraisal of the Environmental Justice Movement*. Cambridge, Mass.: The MIT Press.

Pellow, David, and Lisa Sun-Hee Park. 2002. *The Silicon Valley of Dreams: Environmental Injustice, Immigrant Workers, and the High-Tech Global Economy*. New York: New York University Press.

Pellow, David N., Adam Weinberg, and Allan Schnaiberg. 2001. "The Environmental Justice Movement: Equitable Allocation of the Costs and Benefits of Environmental Management Outcomes." *Social Justice Research* 14:423–439.

Peña, Devon. 1997. *Terror of the Machine: Technology, Work, Gender and Ecology on the U.S.-Mexico Border*. Austin: University of Texas Press.

Pennycook, Frank, Emily Diamand, Andrew Watterson, and Vyvyan Howard. 2004. "Modeling the Dietary Pesticide Exposures of Young Children." *Journal of Occupational and Environmental Health* 10:304–309.

"Pepsi Accused of Flouting Trade Law." 1994. *Times of India,* November 9.

"Pepsi 'Dumping' Plastic Waste." 1994. *Pioneer* (India), November 9.

"Pepsi's Dumping Strategy Raises Public Outcry." 1994. *Asian Age,* November 9.

"Pepsi Failed to Fulfill Promise." 1994. *Hindustan Times,* November 10.

"Pepsi Forays into New Business with Aquafina Bulk Water." 2003. September 29. Domain-b.com http://www.domain-b.com/companies/companies_p/pepsiindia/20030929_new_business.html.

Pesticide Action Network Asia and the Pacific. 2003a. "Delegates Condemn US War on Iraq and Imperialist 'Poisons of Mass Destruction.'" Press release, April 4.

Pesticide Action Network Asia and the Pacific. 2003b. "Pesticide Poisoning Survivors Victory: Agrochemical TNCs Found 'Guilty' of Murder in People's Hearing." January. www.panap.net.

Pesticide Action Network Asia and the Pacific. 2003c. *Empowering Women in Agriculture: Women's Liberation Alive and Kicking.* Malaysia.

Pesticide Action Network Asia and the Pacific. 2003d. *Towards a Pesticides-Free World: Women and Pesticides Poisoning.* Malaysia.

Pesticide Action Network North America. 2003. *Farmworker Women and Pesticides in California's Central Valley.* San Francisco, CA: PANNA.

Pesticide Action Network North America. N.d. "Poison Profits—The G7 Pesticide Industry's Stake in the World Bank." Briefing Paper. San Francisco.

Pesticide Action Network Updates Service. 2002. "U.S. Pesticide Exports Remain High." January 11. http://www.panna.org/resources/panups/panup_20020111.dv.html

Philo, Greg, and David Miller (Eds.). 2001. *Market Killing: What the Free Market Does and What Social Scientists Can Do About It.* New York: Longman.

Piette, Betsey. 1998. "Group Accuses Philadelphia Government of Environmental Racism." *Workers World.* New York. June 25.

Plese, Branimir. 2002. "Racial Segregation in Croation Primary Schools: Romani Students Take Legal Action." *Roma Rights,* no. 3 and 4:129–137.

Plese, Branimir. 2004. "Strasbourg Court Finds Hungary in Breach of Human Rights Standards in a Roma Police Brutality Case." *Roma Rights,* no. 3 and 4:99–100.

Porterfield, Andrew, and David Weir. 1987. "The Export of U.S. Toxic Wastes." *Nation,* October 3.

Powell, J. Stephen. 1984. *Political Difficulties Facing Waste-to-Energy Conversion Plant Siting.* Cerrell Associates for California Waste Management Board. Sacramento, California.

Proceedings from the First National People of Color Environmental Leadership Summit. 1991. Washington, D.C., October.

Project Return to Sender. 1997. "Action Alert: Help Return the Toxic Ash Dumped on Haiti 10 Years Ago to the United States." Essential Action, Washington, D.C.

Prokosch, Mike, and Laura Raymond (Eds.). 2002. *The Global Activist's Manual: Local Ways to Change the World.* New York: Thunder's Mouth Press/Nation Books.

Public Interest Research Group. 1994a. *Toxic Waste Trade: A Primer.* Delhi, India, December.

Public Interest Research Group. 1994b. "The Other Side of Pepsi's Operations in India." Press release, Delhi, November.

Puckett, Jim. 1998. "Something Rotten from Denmark: The Incinerator 'Solution' to Aid Gone Bad in Mozambique." *Multinational Monitor* 19(12):24–26.

Pulido, Laura. 1996. "A Critical Review of the Methodology of Environmental Racism Research." *Antipode* 28:142–159.

Pulido, Laura. 2000. "Rethinking Environmental Racism." *Annals of the Association of American Geographers* 90:12–40.

Quintero-Somaini, A., M. Quirindongo, E. Arevalo, D. Lashof, E. Olson, and G. Solomon. 2004. *Hidden Danger: Environmental Health Threats to the Latino Community.* Natural Resources Defense Council. October. New York.

Raffensperger, Carolyn, and Joel Tickner. 1999. *Protecting Public Health and the Environment: Implementing the Precautionary Principle.* Washington, D.C.: Island Press.

Ramasamy, Arjunan. 2002. "Paraquat: Time to Go! You Can Make the Difference." Statement presented to Syngenta Corporation's Shareholders at the company's annual meeting, April.

Ransom, Pamela. N.d. *Women, Pesticides and Sustainable Agriculture:* Women's Environment and Development Organization. New York.

Raphael, Chad and Ted Smith. 2006. "Extended Producer Responsibility for Electronic Equipment in the USA." In Ted Smith, David Sonnenfeld, and David N. Pellow (Eds.), *Challenging the Chip: Labor Rights and Environmental Justice in the Global Electronics Industry.* Philadelphia: Temple University Press.

Recycling Today. 2004. "Singapore Company Building Electronics Recycling Plant in China." March 23.

Redefining Progress. 2004. *2004 Footprint of Nations.* San Francisco: Redefining Progress.

reEarth. 1992. Newsletter, April-December.

Reeves, M., K. Schafer, and A. Katten. 1999. *Fields of Poison: California Farmworkers and Pesticides.* San Francisco, CA: Pesticide Action Network American Regional Center.

Rendell, Edward G. 1998. Letter to Mary H. Jackson, April 17.

Rengam, Sarojeni, and Jennifer Mourin. 2002. "Paraquat Banned in Malaysia!" *Pesticide Monitor* 2(4–6): 27–29.

Reuther, Christopher. 2002. "Spheres of Influence: Who Pays for e-Junk?" *Environmental Health Perspectives* 110(4):A196–A199.

Revkin, Andrew. 1998. "New York Tries to Clean Up Ash Heap in the Caribbean." *New York Times,* January 15.

Risse, Thomas, Stephen Ropp, and Kathryn Sikkink (Eds.). 1999. *The Power of Human Rights: International Norms and Domestic Change.* Cambridge: Cambridge University Press.

Ritz, Dean (Ed.). 2001. *Defying Corporations, Defining Democracy: A Book of History and Strategy.* Croton-on-Hudson, N.Y.: Apex Press.

Roberts, J. Timmons, and Peter E. Grimes. 2002. "World-System Theory and the Environment." In Riley Dunlap, Frederick Buttel, Peter Dickens, and August Gijswijt (Eds.), *Sociological Theory and the Environment: Classical Foundations, Contemporary Insights.* Lanham, Md.: Rowman and Littlefield.

Roberts, J. Timmons, and Melissa Toffolon-Weiss. 2001. *Chronicles from the Environmental Justice Frontline.* Cambridge: Cambridge University Press.

Rodney, Walter. 1981. *How Europe Underdeveloped Africa.* Washington, D.C.: Howard University Press.

Roediger, David. 1999. *The Wages of Whiteness: Race and the Making of the American Working Class.* London: Verso.

Roediger, David. 2006. "The Retreat from Race and Class." *Monthly Review* 58(3):53–61.

Royte, Elizabeth. 2006. "E-Waste@Large." *New York Times,* January 27.

Reuters News Service. 2005. "Tiny Hong Kong Falling Foul of Electronic Waste." China, March 31.

Russell, Edmund. 2001. *War and Nature: Fighting Humans and Insects with Chemicals from World War I to Silent Spring.* Cambridge: Cambridge University Press.

Sachs, Aaron. 1995. "Eco Justice: Linking Human Rights and the Environment." Worldwatch Paper 127. Washington, D.C.: Worldwatch, December.

Said, Edward. 1979. *Orientalism.* New York: Vintage.

Sample, Ian. 2004. "PCs: The Latest Waste Mountain." *Guardian,* March 8.

Sangaralingam, Mageswari. 2004a. Posting to the ICRT Planning List, March 24.

Sangaralingam, Mageswari. 2004b. Posting to the ICRT Planning List, January 1.

Santa Ana, Otto. 2002. *Brown Tide Rising: Metaphors of Latinos in Contemporary American Public Discourse.* Austin: University of Texas Press.

Sawyer, Jon. 1992. "Haiti Seeks Removal of U.S. Waste." *St. Louis Post Dispatch,* May 12.

Saxena, M. C., M. K. Siddiqui, T. D. Seth, C. R. Krishna Murti, A. K. Bhargava, and D. Kutty. 1981. "Organochlorine Pesticides in Specimens from Women Undergoing Spontaneous Abortion, Premature or full-Term Delivery." *Journal of Analytical Toxicology* 5(1):6–9.

Schaeffer, Robert K. 2003. *Understanding Globalization: The Social Consequences of Political, Economic, and Environmental Change.* Lanham, Md.: Rowman and Littlefield.

Schafer, Kristin. 2001. "Ratifying Global Toxics Treaties: The U.S. Must Provide Leadership." *Foreign Policy in Focus*. Vol. 6, no. 31. September. International Relations Center.

Schatz, Amy. 2003. "Dell Changes Recycle Vendors." *Austin American-Statesman*, July 4.

Schnaiberg, Allan. 1980. *The Environment: From Surplus to Scarcity*. New York: Oxford University Press.

Schnaiberg, Allan, and Kenneth Gould. 2000. *Environment and Society: The Enduring Conflict*. New York: St. Martin's Press.

Schurman, Rachel. 2004. "Fighting 'Frankenfoods': Industry Opportunity Structures and the Efficacy of the Anti-Biotech Movement in Western Europe." *Social Problems* 51: 243–268.

Schwartz, Jerry. 2000. "Philadelphia's Well-Traveled Trash: The Ash That Wouldn't Be Thrown Away." Associated Press, September 3.

Seager, Joni. 1994. *Earth Follies: Coming to Feminist Terms with the Environmental Crisis*. New York: Routledge.

Seippel, Ornulf. 2002. "Modernity, Politics, and the Environment: A Theoretical Perspective." In Riley Dunlap, Frederick Buttel, Peter Dickens, and August Gijswijt (Eds.), *Sociological Theory and the Environment: Classical Foundations, Contemporary Insights*. Lanham, Md.: Rowman and Littlefield.

Sennett, Richard, and Jonathan Cobb. 1993. *The Hidden Injuries of Class*. New York: Norton.

Sherriff, Lucy. 2005. "HP a 'Toxic Tech Giant' Says Greenpeace." http://forms.theregister.co.uk/mail_author/?story_url=/2005/05/23/hp_toxic/.

Shiva, Vandana. 1988. *Staying Alive: Women, Ecology, and Development*. London: Zed Books.

Shiva, Vandana. 1991. *The Violence of the Green Revolution: Third World Agriculture, Ecology, and Politics*. London: Zed Books.

Silicon Valley Toxics Coalition. 2001. *Poison PCs and Toxic TVs: California's Biggest Environmental Crisis That You've Never Heard of*. San Jose, Calif.: SVTC.

Silva, Denise Ferreira. 2007. *Toward a Global Idea of Race*. Minneapolis: University of Minnesota Press.

Singh, Kavalit. 1994a. Letter from PIRG (New Delhi) to Global Response (Boulder, Colorado), November 12.

Singh, Kavalit. 1994b. "An Open Letter to Mr. Christopher A. Sinclair." Public Interest Research Group, Delhi, India, November 7.

Skogly, Sigrun I. 1996. "Moving Human Rights out of Geneva: The Need for a Comprehensive Approach to International Human Rights Law." Paper presented at the International Studies Association Annual Meeting, Washington, D.C.

Skrzycki, Cindy, and Joby Warrick. 2000. "EPA Links Dioxin to Cancer." *Washington Post*, May 17.

Slade, Giles. 2006. *Made to Break: Technology and Obsolescence in America.* Cambridge, Mass.: Harvard University Press.

Smith, Andrea. 2005. *Conquest: Sexual Violence and American Indian Genocide.* Boston: South End Press.

Smith, Jackie, Charles Chatfield, and Ron Pagnucco (Eds.). 1997. *Transnational Social Movements and Global Politics: Solidarity Beyond the State.* Syracuse, N.Y.: Syracuse University Press.

Smith, Jackie, and Hank Johnston (Eds.). 2002. *Globalization and Resistance: Transnational Dimensions of Social Movements.* Lanham, Md.: Rowman and Littlefield.

Smith, Neil. 1996. *New Urban Frontiers: Gentrification and the Revanchist City.* London Routledge.

Smith, Ted, David Sonnenfeld, and David N. Pellow. 2006a. "The Quest for Sustainability and Justice in a High-Tech World." In Ted Smith, David Sonnenfeld, and David N. Pellow (Eds.), *Challenging the Chip: Labor Rights and Environmental Justice in the Global Electronics Industry.* Philadelphia: Temple University Press.

Smith, Ted, David Sonnenfeld, and David N. Pellow (Eds.). 2006b. *Challenging the Chip: Labor Rights and Environmental Justice in the Global Electronics Industry.* Philadelphia: Temple University Press.

Snow, David, and Robert Benford. 1992. "Master Frames and Cycles of Protest." In Aldon Morris and Carol Mueller (Eds.), *Frontiers in Social Movement Theory.* New Haven, Conn.: Yale University Press.

Sogge, David. N.d. "A Short History of the Mozambican Civil Society."

Sonnenfeld, David. 2000. "Contradictions of Ecological Modernisation: Pulp and Paper Manufacturing in South-east Asia." *Environmental Politics* 9(1): 235–256.

Spaargaren, Gert. 1997a. *The Ecological Modernization of Production and Consumption.* Wageningen: Wageningen Agricultural University.

Spaargaren, Gert. 1997b. "The Ecological Modernization of Production and Consumption." Ph.D. dissertation, Landbouw University, The Netherlands.

Spaargaren, Gert, and Arthur P. J. Mol. 1992. "Sociology, Environment, and Modernity: Ecological Modernization as a Theory of Social Change." *Society and Natural Resources* 5(4):323–344.

Spivak, Gayatri Chakravorty. 1999. *A Critique of Postcolonial Reason: Toward a History of the Vanishing Present.* Cambridge, Mass.: Harvard University Press.

Sri Lankan Ministry of Health. 1997. *Annual Health Bulletin, Sri Lanka 1995.* Colombo: Ministry of Health.

Starr, Amory. 2000. *Naming the Enemy: Anti-Corporate Movements Confront Globalization.* London: Zed Books.

Steingraber, Sandra. 1997. *Living Downstream: An Ecologist Looks at Cancer and the Environment.* Reading, Mass.: Addison-Wesley.

Stiffler, Lisa. 2004. "Effort Will Target Harmful Fire Retardant." *Seattle Post-Intelligencer,* January 29.

Stiglitz, George. 2003. *Globalization and Its Discontents.* New York: Norton.

Stiles, Kendall. 1998. "Civil Society Empowerment and Multilateral Donors: International Institutions and New International Norms." *Global Governance* 4(2):199–216.

Summers, Lawrence. 1991. Internal memo. World Bank. December 12.

Szasz, Andrew. 1994. *Ecopopulism: Toxic Waste and the Movement for Environmental Justice.* Minneapolis: University of Minnesota Press.

Szasz, Andrew, and Michael Meuser. 1997. "Environmental Inequalities: Literature Review and Proposals for New Directions in Research and Theory." *Current Sociology* 45:99–120.

Sze, Julie. 2003. "Noxious New York." Ph.D. dissertation, New York University.

Szendrey, Orsolya. 2002. "Hungarian Villagers Enforce Mob Justice Solution to Prevent Roma from Moving in." *Roma Rights,* no. 3 and 4: 70–72.

Takaki, Ronald. 1998. *Strangers from a Different Shore: A History of Asian Americans.* Boston: Back Bay Books.

Tangri, Neil . 2003. *Waste Incineration: A Dying Technology.* Washington, D.C.: Essential Action. July.

Tarr, Joel. 1996. *The Search for the Ultimate Sink: Urban Pollution in Historical Perspective.* Akron, Ohio: University of Akron Press.

Tarrow, Sidney (Ed.). 1998. *Power in Movement: Social Movements and Contentious Politics.* Cambridge: Cambridge University Press.

Taylor, Bron (Ed.). 1995. *Ecological Resistance Movements: The Global Emergence of Radical and Popular Environmentalism.* Albany: State University of New York Press.

Taylor, Dorceta. 1997. "American Environmentalism: The Role of Race, Class and Gender in Shaping Activism, 1820–1995." *Race, Gender and Class* 5:16–62.

Tenaganita and PANAP. 2002. *Poisoned and Silenced: A Study of Pesticide Poisoning in the Plantation.* Malaysia, March.

Texaco. 1994. "Texaco Announces Settlement in Class Action Lawsuit." White Plains, N.Y., November 15. http://www.chevron.com/news/archive/Texaco_press/1996/pr11_15.asp.

Texas Campaign for the Environment. 2003a. "An Open Letter to Michael Dell." July 14.

Texas Campaign for the Environment. 2003b. "Dell Accepts E-waste Collected by Activists." Computer TakeBack Campaign press release,. Austin, Texas, July 19.

Thompson, Eulalee. N.d. *Eco-Report.* London: Panos Institute.

Tiemann, M. 1998. "Waste Trade and the Basel Convention: Background and Update." *Congressional Research Service Report for Congress.* Committee for the National Institute for the Environment. Washington, D.C.

Times of India Online. 2004. "NGO Sounds E-waste Alert." March 16.

Todorov, Tzvetan. 1984. *The Conquest of America*. New York: Harper.

Toynbee, Arnold. 1934. *A Study in History.* New York: Oxford University Press.

Tolba, Mustafa K. (Ed.). 1988. *Evolving Environmental Perceptions: From Stockholm to Nairobi.* Nairobi: United Nations Environment Programme.

Towles, Robert. 1993. Letter to Sandra Marqardt, "Re: Bahamas, Great Abaco Sugar Mill." March 29.

"Toxic Chemicals Leaving Abaco." 1994. *Abaconian* 2(11).

Toxics Link India. 2003. "Fact File on Waste in the Wireless World." http://www.toxicslink.org/. India.

Toxics Link India. 2002. "Killing Fields of Warangal." New Delhi, March.

Toxics Link India. 2003. *Scrapping the Hi-Tech Myth: Computer Waste in India.* New Delhi, India

Toxic Trail (documentary film). 2001. Television Trust for the Environment. http://www.toxictrail.org/ToxicTrail-Summary.htm.

"Toxic Waste Leaves Abaco." 1995. *Abaconian* 3(1).

Turé, Kame [Carmichael, Stokely], and Charles V. Hamilton. [1968] 1992. *Black Power: The Politics of Liberation.* New York: Vintage.

United Church of Christ. 1987. *Toxic Wastes and Race in the United States.* New York: UCC Commission for Racial Justice.

United for a Fair Economy. 2002. *The Global Activist's Manual: Local Ways to Change the World.* New York: Nation Books.

United Nations. 1973. *Report of the United Nations Conference on the Human Environment, Stockholm, June 5–16.* New York: United Nations.

United Nations Conference on Trade and Development. 1998. *Least Developed Countries 1998 Report: Overview.* Geneva: UNCTAD.

United Nations Development Program. 1999. *Human Development Report 1999.* Geneva: UNDP.

United Nations. 2005. *Ecosystems and Human Well-Being (*Millennium Ecosystem Assessment). Washington, D.C.: Island Press and the United Nations.

United Nations Population Fund. 2001. *The State of World Population: Women and the Environment.* New York: United Nations.

U.S. Department of Commerce. 2002. *A Nation Online: How Americans are Expanding Their Use of the Internet.* Washington, D.C.: Economics and Statistics Administration and National Telecommunications and Information Administration, February.

U.S. Newswire. 2001. "U.N. Human Rights Investigator Deems U.S. Export of Banned Pesticides 'Immoral.'" San Francisco, December 17.

van der Heijden, Hein-Anton. 2006. "Globalization, Environmental Movements, and International Political Opportunity Structures." *Organization and Environment* 19(1): 28–45.

Veldhuizen, Hennie, and Bob Sippel. 1994. "Mining Discarded Electronics." *UNEP Industry and Environment*. July-September, 7–11.

Vidal, John. 2004. "They Call This Recycling, But It's Really Dumping by Another Name." *Guardian,* September 21.

Vinayak, Chaturvedi. (Ed.). 2000. *Mapping Subaltern Studies and the Postcolonial.* London: Verso.

Vo, Linda. 2004. *Mobilizing an Asian American Community.* Philadelphia: Temple University Press.

Wallach, Lori, and Patrick Woodall. 2004. *Whose Trade Organization? A Comprehensive Guide to the* WTO. New York: New Press.

Wankhade, Kishore. 2004. "British Environment Agency Report Reveals 23,000 Tons of E-Waste Being Illegally Exported to Developing Nations, including India." New Delhi: Toxics Link, December 17. http://www.toxicslink.org/mediapr-view.php?pressrelnum=18.

Ward, Mike, 1987. "State Board Denies Using Siting Report; Study Identifies Least Likely Incinerator Foes." *Los Angeles Times,* July 16.

Wargo, John. 1998. *Our Children's Toxic Legacy: How Science and Law Fail to Protect Us from Pesticides.* 2nd ed. New Haven, Conn.: Yale University Press.

"Waste Dumping Grounds of the World." 2005. June 28. http://www.thesouthasian.org/archives/000396.html.

"Waste Not, Want Not." 2002. Report by Prime Minister Tony Blair's Strategy Unit. London.

Weale, Albert. 1992. *The New Politics of Pollution.* Manchester: Manchester University Press.

Weinberg, Adam, David N. Pellow, and Allan Schnaiberg. 2000. *Urban Recycling and the Search for Sustainable Community Development.* Princeton, N.J.: Princeton University Press.

Weir, David. 1987. *The Bhopal Syndrome: Pesticides, Environment, and Health.* San Francisco: Sierra Club Books.

Weir, David, and Mark Schapiro. 1981. *Circle of Poison: Pesticides and People in a Hungry World.* Oakland, Calif.: World Health Organization.

Wesseling, C., R. McConnell, T. Partanen, and C. Hogstedt. 1997. "Agricultural Pesticide Use in Developing Countries: Health Effects and Research Needs." *International Journal of Health Services* 27:273–308.

West, Cornel. 1997. "The Ignoble Paradox of Western Modernity." In Madeleine Burnside and Rosemarie Robotham (Eds.), *Spirits of the Passage: The Transatlantic Slave Trade in the Seventeenth Century.* New York: Simon and Schuster.

West, Cornel. 1999. *Cornel West Reader.* New York: Basic Civitas.

Westra, Laura. 1998. "Development and Environmental Racism: The Case of Ken Saro-Wiwa and the Ogoni." *Race, Gender, and Class* 6:152–162.

Williams, Martyn. 2006. "Computex: Greenpeace Protest on Video." IDG News Service. Taipei, Taiwan.

Williams, Orrin. 2000. "Environmental Racism and the Destruction of African Civilization." Presentation at the University of Colorado, Boulder, Colo., Fall.

Williams, Patricia. 1997. *Seeing a Colorblind Future: The Paradox of Race*. New York: Noonday Press.

Williams, Richard. 1990. *Hierarchical Structures and Social Value*. Cambridge: Cambridge University Press.

Wilson, William Julius. 1990. *The Truly Disadvantaged: The Inner City, the Underclass, and Public Policy*. Chicago: University of Chicago Press.

Winant, Howard. 2001. *The World Is a Ghetto: Race and Democracy since World War II*. New York: Basic Books.

Wines, Michael. 2006. "A Trial Ends in Mozambique, But Many Questions Hover." *New York Times*, January 1.

Wise, Tim. 2001. "School Shootings and White Denial." March 6. www.alternet.org.

Wise, Tim. 2005. *White Like Me: Reflections on Race from a Privileged Son*. Brooklyn, N.Y.: Soft Skull Press.

Witness for Peace-Mid Atlantic. 1998. "Citizens Group Travels to Haiti to Investigate Effects of Philadelphia's Toxic Waste Dump There." Press release, February 27.

World Commission on Environment and Development. 1987. *Our Common Future*. New York: Oxford University Press.

World Health Organization. 1990. *Public Health Impact of Pesticides Used in Agriculture*. Geneva. WHO and UNEP

World Health Organization. 1992. *Our Planet, Our Health*. Geneva: WHO.

Wolff, M. S., et al. 1993. "Blood Levels of Organochlorine Residues and Risk of Breast Cancer." *Journal of the National Cancer Institute* 85:648–652.

Wolfowitz, Paul. 1996. "Good Governance and Development: A Time for Action." Speech in Jakarta, Indonesia, April 11.

Wood, Nicholas. 2006. "Displaced Gypsies at Risk from Lead in Kosovo Camps." *New York Times*, February 5.

Wright, Beverly. 2005. "Living and Dying in Louisiana's Cancer Alley." In Robert Bullard (Ed.), *The Quest for Environmental Justice: Human Rights and the Politics of Pollution*. San Francisco: Sierra Club Books.

Xing, Danwen. 2003. "How Do We Look?" *New York Times*, June 21, A27.

Yang, Tseming. 2002. "International Environmental Protection: Human Rights and the North-South Divide." In Kathryn Mutz, Gary Bryner, and Douglas Kenney (Eds.), *Justice and Natural Resources: Concepts, Strategies, and Applications*. San Francisco: Island Press.

Yatim, A. Hafiz. 2004. "Malacca Rejects RM63m Recycling Plant Project." *New Straits Time*, January 1.

Yearley, Steven. 1996. *Sociology, Environmentalism, Globalization*. Thousand Oaks, Calif.: Sage.

Yearley, Steven. 2002. "The Social Construction of Environmental Problems: A Theoretical Review and Some Not-Very-Herculean Labors." In Riley Dunlap, Frederick Buttel, Peter Dickens, and August Gijswijt (Eds.), *Sociological Theory and the Environment: Classical Foundations, Contemporary Insights*. Lanham, Md.: Rowman and Littlefield.

Zavella, Patricia. 1987. *Women's Work and Chicano Families: Cannery Workers of the Santa Clara Valley*. Ithaca, N.Y.: Cornell University Press.

Index